APPROXIMATE ANALYTICAL METHODS FOR SOLVING ORDINARY DIFFERENTIAL EQUATIONS

APPROXIMATE ANALYTICAL METHODS FOR SOLVING ORDINARY DIFFERENTIAL EQUATIONS

T.S.L. RADHIKA
T.K.V. IYENGAR
T. RAJA RANI

CRC Press
Taylor & Francis Group
Boca Raton London New York

CRC Press is an imprint of the
Taylor & Francis Group, an **informa** business
A CHAPMAN & HALL BOOK

CRC Press
Taylor & Francis Group
6000 Broken Sound Parkway NW, Suite 300
Boca Raton, FL 33487-2742

First issued in paperback 2019

ISBN-13: 978-1-4665-8815-8 (hbk)
ISBN-13: 978-0-367-37812-7 (pbk)

Library of Congress Cataloging-in-Publication Data

Radhika, T. S. L., 1974-
Approximate analytical mathods for solving ordinary differential equations / T.S.L. Radhika, T.K.V. Iyengar, and T. Raja Rani.
pages cm
"A CRC title."
Includes bibliographical references and index.
ISBN 978-1-4665-8815-8
1. Differential equations. I. Iyengar, T. K. V. II. Rani, T. Raja, 1972- III. Title.

QA372.R16 2015
515'.352--dc23 2014021805

Visit the Taylor & Francis Web site at
http://www.taylorandfrancis.com

and the CRC Press Web site at
http://www.crcpress.com

Contents

Preface

It is needless to say that differential equations play an important role in modeling many physical, engineering, technological, and biological processes. The differential equations in these diverse contexts may not be directly solvable by the usual elementary methods and hence in general do not have exact or closed-form solutions. In all such cases, researchers have tried to obtain either analytical approximate solutions or numerical approximate solutions. With the available high-speed computers and excellent computational algorithms, considerable advancement has been made in obtaining good numerical solutions. However, there have been trials to obtain approximate analytical solutions, and several approximate analytical methods have been developed to cater to the needs that have arisen and with a view to obtain "better-and-better" solutions. The methods range from the classical series solution method to the diverse perturbation methods and from the pioneering asymptotic methods to the recent ingenious homotopy methods.

This book aims to present some important approximate methods for solving ordinary differential equations and provides a number of illustrations. While teaching some related courses, we felt the need for a book of this type because there is no single book with all the available approximate methods for solving ordinary differential equations. At present, a student or a researcher interested in understanding the state of the art has to wade through several books and research articles to grasp the diverse methods. This book covers both the well-established techniques and the recently developed procedures along with detailed examples elucidating the applications of these methods to solve related real-world problems. It aims to give a complete description of the methods considered and discusses them through illustrative examples without going into several of the rigorous mathematical aspects.

Chapter 1 is introductory. We explain briefly the methods chosen for discussion in the present work.

Chapter 2 introduces the classical method of solving differential equations through the power series method. In fact, this method has been the basis for the introduction and the development of various special functions found in the literature. We explain and illustrate the method with a number of examples and proceed to describe the Taylor series method.

Chapter 3 deals with the asymptotic methods, which can be used to find asymptotic solutions to the differential equations that are valid for large values of the independent variable and in other cases as well.

The introduction of perturbation methods, which constitutes one of the top ten progresses of theoretical and applied mechanics of the 20th century, is the focus of Chapter 4. Attention is drawn to some research articles in which the perturbation methods are used successfully in understanding some physical phenomena whose mathematical formulation involves a so-called perturbation parameter.

Chapter 5 focuses on a special asymptotic technique called the multiple-scale technique for solving the problems whose solution cannot be completely described on a single timescale.

Chapter 6 describes an important asymptotic method called the WKB (for its developers, Wentzel, Kramers, and Brillown) method that helps construct solutions to problems that oscillate rapidly and problems for which there is a sudden change in the behavior of the solution function at a point in the interval of interest.

Chapter 7 deals with some nonperturbation methods, such as the Adomian decomposition method, delta expansion method, and others, that were developed during the last two decades and can provide solutions to a much wider class of problems.

Chapter 8 presents the most recent analytical methods developed, which are based on the concept of homotopy of topology and were initiated by Liao. The methods are the homotopy analysis method, homotopy perturbation method, and optimal homotopy asymptotic method.

Our principal aim is to present and explain the methods with emphasis on problem solving. Many illustrations are presented in each chapter.

The content of this book was drawn from diverse sources, which are cited in each chapter. Further, attention is drawn to some research articles that discuss use of the methods. We are grateful to the authors of all the works cited.

We believe this book will serve as a handbook not only for mathematicians and engineers but also for biologists, physicists, and economists. The book presupposes knowledge of advanced calculus and an elementary course on differential equations.

Acknowledgments

The authors wish to convey their sincere gratitude to their respective institutions, BITS PILANI- Hyderabad campus, India; NIT, Warangal, India; and Military Technological College, Muscat, Oman for providing resources and support while developing this book. They are indebted to their colleagues for their interest and constant encouragement.

The authors are grateful to their family members for their unflinching support and endurance without which this book would not have been possible.

T Raja Rani wishes to place on record her sincere thanks to Prof CNB Rao for his mentorship and guidance while writing this book.

1

Introduction

Many real-world problems involving change of variables with reference to spatial variables or time in science and engineering are modeled by researchers through differential equations with imposed boundary or initial conditions or both. Even if the governing equations are partial differential equations, solutions of the problems, in general, involve solutions of ordinary differential equations. As nonlinearity is the law of nature, the models involve solving nonlinear ordinary differential equations. Here again, the nonlinearity may come under weak nonlinearity or strong nonlinearity. Thanks to the advancement of high-speed computers and remarkable computer algorithms, obtaining reasonably good numerical solutions to linear problems is not now an uphill task. However, we come across some problems that are fairly simple to understand and visualize but involve solving highly nonlinear differential equations that cannot be solved exactly/analytically.

We come across problems of the following types in the literature:

1. Problems that involve differential equations exactly representing the problems realistically for which exact solutions can be obtained.
2. Problems that involve differential equations arising because of some simplifying assumptions and thus are approximate representations of the original problems but can be solved exactly.
3. Problems that involve differential equations exactly representing the problems under consideration but for which solutions cannot be obtained exactly so we may have to try to obtain an approximate solution.
4. Problems that involve differential equations arising because of some simplifying assumptions and thus are approximate representations of the original problems and also cannot be solved exactly.

Problems from various branches of science and engineering, unfortunately, do not always fall under category 1. Problems of category 2 are all right, but the solutions of the problems are not the exact solutions of the original problems.

When we formulate the real-world problems either exactly or approximately, suppose we are unable to obtain exact solutions, as in categories 3 and 4; then, we necessarily have to resort to obtaining approximate solutions of the problems formulated.

The approximate solutions may be

1. analytical approximate solutions
2. numerical approximate solutions

As mentioned, because of the availability of high-speed computers and techniques in programming, a good number of numerical techniques have been developed to solve problems in differential equations with considerable complexity and nonlinearity. In spite of this, the quest for analytical solutions, whether exact or approximate, has always existed, and a number of newer methods proposed by ingenious workers are making their presence felt.

Consider a differential equation of the form

$$L(u,x,\varepsilon)=0 \quad \text{or} \quad L(u,x)=0 \tag{1.1}$$

where x is an independent variable, u is the dependent variable, ε is a small parameter, and L is a differential operator.

Let

$$B(u,\varepsilon)=0 \quad \text{or} \quad B(u)=0 \tag{1.2}$$

be a condition to be satisfied by u. This may be taken as a boundary condition or an initial condition.

If we can, by some method, find $u(x)$, which satisfies Equations (1.1) and (1.2) exactly, that would be ideal. In this context, note that for an equation of the form of Equation (1.1) it is far more tractable to obtain an exact analytical solution if it is linear than if it is nonlinear. As mentioned, when we cannot obtain an exact analytical solution, the question is whether it is possible to obtain an analytical solution that is approximate. Over many decades, several analytical procedures, mostly approximate analytical methods, were developed by researchers to partially answer this question. Some of the popular methods in this direction are the following:

1. Power series solution method
2. Asymptotic series solution method
3. Perturbation methods
4. Nonperturbation methods such as Lyapunov's artificial parameter method, δ expansion method, and so on
5. Adomian decomposition method
6. Homotopy methods

The series solution method is classical in the sense that it led to a vast field and development in the form of special functions, such as Legendre polynomials, Bessel functions, Hermite polynomials, wave functions, and so on. The series solutions constructed using the power series method are all convergent in nature, although in some cases the region of convergence is small. But, the major drawback is that this method fails to provide solutions to the equations at so-called irregular singular points and the solutions that are valid for the point at infinity. An offshoot of the power series method is the asymptotic method, which takes care of the asymptotic behavior of the solutions obtained. Chapter 2 explains the power series method and presents several illustrations involving linear and nonlinear equations.

The asymptotic method also provides a solution to the problem in terms of an infinite series, but it need not always be a convergent series. The characteristic feature of these solutions is that, in spite of their divergent nature, there will be a particular partial sum that provides the best approximation to the solution function of the differential equation considered. To construct solutions valid for large values of the independent variable, a technique called the Lindstedt-Poincaré technique, also known as the method of stretched coordinates, is explained (the basic ideas for this emanated in the late nineteenth century). Chapter 3 provides the basics of the asymptotic method together with the various techniques for finding asymptotic solutions to different classes of problems.

A differential equation governing a physical phenomenon, whether linear or nonlinear, can involve certain parameters. Students of fluid mechanics, elasticity, structural mechanics, quantum mechanics, and diverse other fields are well aware of this aspect. By changing the parameters slightly (or, in other words, by perturbing the parameters), the solution may be changed slightly or otherwise. Problems in this class are termed perturbation problems. Again, there are broadly two varieties of perturbation problems: regular and singular.

One of the earliest techniques has been to express the unknown function or functions of the problem to be solved as a power series expansion in terms of this parameter (referred to as the perturbation parameter) and obtain the successive terms of the expansion with care. This is known as the perturbation technique. Using this technique and diverse improvements on it, several problems from various fields have been solved. In fact, the perturbation technique for solving problems is supposed to be one of the greatest advancements of the twentieth century. However, note that this method is aided or supplemented by a number of asymptotic methods, such as the boundary-layer method, multiple-scale method, WKB (for its developers, Gregor Wentzel, Hendrik Kramers, and Léon Brillown) method, and so on. Chapter 4 deals with the perturbation techniques for solving the so-called regular perturbation problems and details the boundary-layer method, also known as the matched asymptotic technique, for providing solutions to singularly perturbed problems. Chapters 5 and 6 throw light on two popular asymptotic techniques: the multiple-scale method and the WKB method, respectively.

In all most all the methods described, when the problem involves a linear or nonlinear differential equation, broadly we develop a sequence of linear subproblems that are again differential equations, which we successively solve and take the sum as the solution of the original problem.

Note that not every problem has a perturbation parameter involved; also, even if one exists, the subproblems constructed may become so complicated that only a few of them have a solution. Hence, to solve this category of problems, researchers have developed the so-called nonperturbation methods. Among them are Lyapunov's artificial parameter method and the δ expansion method. In these methods, again the solution to the problem is assumed to be an infinite series in terms of a small parameter that is introduced into the problem. A series of subproblems that are linear ordinary differential equations are then constructed. The advantage that is gained by introducing this so-called artificial parameter is that it can be placed in the equation in any term of our choice so that it is almost possible to find solutions to all the subproblems constructed.

Another nonperturbation technique considered to be a powerful analytical technique for solving ordinary differential equations is the Adomian decomposition method. It can be used to solve strongly nonlinear differential equations as well. Unlike the perturbation techniques and the other nonperturbation techniques, it needs neither a perturbation parameter nor an artificial parameter. The solution here is expressed in terms of "Adomian polynomials," which can be easily computed. In fact, codes are available for the computation of these polynomials in software such as Mathematica®. The solution constructed using this decomposition technique is in general convergent. Chapter 7 of the book provides an insight into these three nonperturbation techniques.

The last chapter of the book details a recent "nicely developed" homotopy analysis method proposed by Liao that overcomes the limitations of the earlier methods touched on in this book. It is also a semianalytical technique that is best suited for nonlinear ordinary and partial differential equations. It employs the concept of "homotopy" of topology to generate a convergent solution for not only weakly but also strongly nonlinear problems, going "beyond" the limitations of the perturbation/nonperturbation techniques. In fact, this method unifies Lyapunov's artificial parameter method, the δ expansion method, and the Adomian decomposition method.

This book gives a complete description of the proposed methods and illustrates them through examples. The methods are introduced and explained without going into several of the rigorous mathematical aspects. After basic exposure to the working procedures, references suggested at the end of each chapter provide greater insight. The book also includes a list of many research articles related to the applications of the techniques in various fields. This list, of course, is not exhaustive but is sufficient for an interested student.

2
Power Series Method

Introduction

Solutions to differential equations that can be expressed exactly in terms of elementary functions like polynomials, rational functions, trigonometric functions, logarithmic functions, and so on are referred to as exact solutions. For example, the differential equation $y' + 3y = 0$ has $y = Ae^{-3x}$ as its solution, and this is an exact solution. But, it may not always be possible to find such solutions to differential equations that arise in several applications. For the equation $y'' - xy = 0$ (in literature, this equation is referred to as Airy's differential equation), which is used to model the diffraction of light, it is not possible to find $y(x)$ that exactly satisfies the equation using the classical methods for solving ordinary differential equations (ODEs) (you may verify this). Hence, for such equations, we seek an approximate solution in terms of an infinite series that is arranged as powers of the independent variable. The series solution thus obtained is called a power series solution and is always a convergent series. The method of finding this form of solution is termed the power series method. This chapter describes two methods:

1. Algebraic or method of undetermined coefficients
2. Taylor series method

for finding power series solutions to ODEs.

To begin, we consider some important definitions in this section. Sections 2.2 and 2.3 introduce the algebraic method or the method of undetermined coefficients to find solutions to differential equations at so-called ordinary points and singular points. Section 2.4 gives a brief remark on the series solutions at irregular singular points (IRSPs). The Taylor series method is discussed in Section 2.5. Also provided are a list of articles and books for ready reference together with a list of research articles that provide greater insight on the applications of the method discussed.

Definition

An infinite series of the form

$$\sum_{n=0}^{\infty} a_n (x - x_0)^n \tag{2.1}$$

where a_n's are constants, is called a power series about the point (centered) $x = x_0$.

In particular, if $x_0 = 0$, Equation (2.1) is said to be a power series centered at the origin. ■

Definition: Interval of Convergence and Radius of Convergence

If there exists a positive real number R such that the series Equation (2.1) is convergent for $|x - x_0| < R$ and divergent for $|x - x_0| > R$, then R is called the radius of convergence of the power series, and the interval $(-R, R)$ is called the interval of the convergence of the series Equation (2.1). In fact, it will be the interval of absolute convergence of the power series. ■

Note: The radius of convergence R can be calculated by the formula

$$R = \lim_{n \to \infty} \left| \frac{a_n}{a_{n+1}} \right|.$$

This radius of convergence R can be zero, finite, or infinite as well.

For rigorous proof, refer to some standard works on real analysis (Apostol, 1974; Rudin, 1976).

Definition

If $f(x)$ has a power series expansion of the form $\sum_{n=0}^{\infty} a_n (x - x_0)^n$ with radius of convergence R, then $f(x)$ is said to be analytic at $x = x_0$ with R as the radius of analyticity. ■

Let us look at some examples for which we calculate the radius of convergence of some power series.

Example 2.1

For the series

$$\sum_{n=0}^{\infty}\frac{(-3)^n}{n\ 5^{n+1}}(x-3)^n,\quad R=\lim_{n\to\infty}\left|\frac{(-3)^n.(n+1).5^{n+2}}{(-3)^{n+1}.n.5^{n+1}}\right|=\frac{5}{3}$$

Hence, the series has a radius of convergence $R=\frac{5}{3}$, and the interval of convergence of this power series is $(-5/3,5/3)$.

Example 2.2

The radius of convergence of $\sum_{n=0}^{\infty}\frac{x^n}{n!}$ is

$$R=\lim_{n\to\infty}\left|\frac{n+1!}{n!}\right|=\infty$$

Hence, from the definition of the interval of convergence, we can say that this series is convergent for all x.

Example 2.3

For the series

$$\sum_{n=0}^{\infty}n!x^n,\quad R=\lim_{n\to\infty}\left|\frac{n!}{n+1!}\right|=0,$$

hence this series is convergent only at $x=0$ and nowhere else.

The series in Example 2.3 cannot be treated as a valid solution series as it is convergent only at a single point, namely, the origin. So, we shall construct a series solution to ODEs that has a finite and nonzero radius of convergence. We now begin with the method of undetermined coefficients or the algebraic method for finding a power series solution to the differential equations.

Algebraic Method (Method of Undetermined Coefficients)

Introduction and Some Important Definitions

Definition: Ordinary Point and Singular Point of a Differential Equation
Consider the differential equation given by

$$y' + P(x)y = 0, \quad y(x_0) = y_0 \tag{2.2}$$

If $P(x)$ is analytic at x_0 with R as the radius of analyticity, then Equation (2.2) has a power series solution at $x = x_0$ given by $y(x) = \sum_{n=0}^{\infty} a_n (x - x_0)^n$. Such a point x_0 is called an ordinary point of the differential Equation (2.2). As R is the radius of analyticity of $P(x)$, the power series solution is valid in $|x - x_0| < R$.

We may note a point here that, usually, the value of R is nonzero; thus, we always obtain a power series solution that is valid in some interval whether it is small or large.

Otherwise, x_0 is called a singular point of Equation (2.2).
Let us begin with a simple example that illustrates these above concepts. ■

Example 2.4

Consider the equation

$$y' + xy = 0, \quad y(0) = 1.$$

Here, $P(x) = x$, which is analytic at $x = 0$. Thus, $x = 0$ is an ordinary point of this differential equation. Hence, this equation has a power series solution about the origin given by $y = \sum_{n=0}^{\infty} a_n x^n$, where a_n's are constants to be determined.

Further, as $R = \infty$, this series solution is valid for $|x| < \infty$.

Example 2.5

In a second example, we consider the differential equation

$$y' - \frac{y}{x-1} = 0, \quad y(0) = 1.$$

Here, $P(x) = -\frac{1}{x-1}$. Hence, $x = 0$ is an ordinary point, and the series solution centered at the origin for the differential equation is $y = \sum_{n=0}^{\infty} a_n x^n$. This power series solution is valid only for $|x| < 1$ (you may check this).

You may note that $P(x)$ is not analytic at $x = 1$; thus, $x = 1$ is a singular point of the equation.

Example 2.6

Now, consider the equation

$$xy' - y = 0.$$

This can be written as

$$y' - \frac{y}{x} = 0.$$

Here, $P(x) = -\frac{1}{x}$, which is not analytic at $x = 0$.

Hence, $x = 0$ is a singular point of the differential equation.

At this point, let us not try to construct a series solution at this so-called singular point of the differential equation. We illustrate the method of constructing solutions to equations at such a point further in the book.

Now, we turn our attention to a second-order, linear, and homogeneous differential equation of the form

$$y'' + P(x)y' + Q(x)y = 0, \quad y(x_0) = a_0, \quad y'(x_0) = a_1 \tag{2.3}$$

Equation (2.2) is referred to as an initial value problem (IVP).

Now, the power series solution to this problem is defined: If both $P(x)$ and $Q(x)$ are analytic at $x = x_0$, then Equation (2.3) has a power series solution of the form [as in the case of Equation (2.2)] $y(x) = \sum_{n=0}^{\infty} a_n (x - x_0)^n$, and this solution is valid on $|x - x_0| < R$, where R is the radius of analyticity of both $P(x)$ and $Q(x)$ about $x = x_0$.

Let us take $P(x) = \frac{1}{x-3}$ and $x_0 = 0$; then, the radius of analyticity of $P(x)$ is $R = |0 - 3| = 3$. If $Q(x) = \frac{1}{x-2}$, then its radius of analyticity about the same point $x_0 = 0$ would be $R = |0 - 2| = 2$. Thus, the radius of analyticity of $P(x)$ and $Q(x)$ about $x = 0$ is 2. Hence, the power series solution about the origin of the Equation (2.2) with $P(x)$ and $Q(x)$ taken as presented has a valid solution in $|x - 0| < 2$.

Some special differential equations are now shown and the concepts defined are applied. These differential equations occur frequently in many applications in physics and engineering, as mentioned in the following material.

Example 2.7

First, consider the equation

$$(1-x^2)y''-2xy'+n(n+1)y=0.$$

This is called Legendre's differential equation. It has a power series solution about $x=0$ that is valid in the interval $|x|<1$. For certain values of n, the infinite series solution breaks down into polynomials. These are called Legendre polynomials, which occur as solutions to wave equations, Schrödinger's equation, and so on.

For more details, refer to books on special functions (Bell, 2004; Zill, 2008).

Example 2.8

Consider the differential equation

$$(1-x^2)y''-xy'+p^2y=0,$$

where p is a real constant. In literature, this equation is called Chebyshev's equation, named after the Russian mathematician Pafnuty Chebyshev. It has $x=0$ as its ordinary point and has a power series solution that is valid in the interval $|x|<1$. Even for this equation, for some specific values of p, the infinite series solutions break down into polynomials, which are called Chebyshev polynomials. These polynomials occur as solutions to the differential equations in the study of electric circuit theory (Bell, 2004).

Example 2.9

Airy's equation $y''-xy=0$ has no singular points; hence, the power series solution about $x=0$ is valid for all x. In fact, power series solutions exist at all real x.

In literature, this equation is also termed the Stokes equation. Its solutions are called Airy functions. The Airy function occurs as the solution to Schrödinger's equation for a particle confined within a triangular potential well and for a particle in a one-dimensional constant force field. Also, this equation has a unique feature that, for $x < 0$, the solutions have an oscillating nature, whereas for $x > 0$, the solutions are of an exponential nature. The point $x = 0$ is called a turning point. Solutions to these equations with turning points can be found using another method called, WKB (for its developers, Gregor Wentzel, Hendrik Kramers, and Léon Brillown), which is detailed in Chapter 6.

There are a few more standard differential equations that have wide range of applications. For more details, refer to the work of Bell (2004).

To this point, we have considered first- and second-order differential equations. Now, we apply the concept of ordinary and singular points to higher-order equations. The definitions given previously can be extended with ease for the higher-order equations, as shown next.

Example 2.10

The differential equation

$$x^3 y''' + 3x^2 y'' - 2xy' + 2y = 0$$

has $x = 0$ as its singular point.

Now, we discuss the method for solving a first-order differential equation using the series solution method at an ordinary point. This method can be extended for finding series solutions to higher-order equations as well.

Solution at Ordinary Point of an Ordinary Differential Equation

Method for First-Order Equations

Consider the first-order differential equation

$$y' + P(x)y = 0$$

Step 1: Rewrite the given equation in the form

$$A(x)y' + B(x)y = 0$$

Step 2: If not already specified, choose x_0 such that $A(x_0)$ is nonzero so that x_0 is an ordinary point.

Step 3: Assume a power series solution

$$y(x) = \sum_{n=0}^{\infty} a_n (x - x_0)^n$$

that is convergent in $|x - x_0| < R$, where R is the radius of convergence defined previously.

Step 4: Find

$$y'(x) = \sum_{n=1}^{\infty} n a_n (x - x_0)^{n-1}$$

Step 5: Substitute $y(x)$ and $y'(x)$ in the given differential equation and collect the coefficients of $(x - x_0)^n$ for each n. Then, we obtain a relation among a_n's. Usually, these will be recurrence relations that will lead, in general, to the determination of a_n's.

Step 6: Substitute the values of a_n's in $y(x)$ to obtain the desired series solution to the given equation.

This method can be applied to solve nonlinear differential equations, perhaps with the limitation that the definitions for determining the radius of convergence for the solution cannot be applied for those series solutions.

Let us now see some examples.

Example 2.11

Solve $y' - 2xy = 0$ with $y(0) = 1$:

Solution: Note that $P(x) = -2x$, which is analytic for all real x, and x_0 is given to be 0 (in view of the given initial condition).

Applying the method described previously, we take $y(x) = \sum_{n=0}^{\infty} a_n x^n$, and substituting in the given equation, we have

$$\sum_{n=1}^{\infty} n a_n x^{n-1} - 2\sum_{n=0}^{\infty} a_n x^{n+1} = 0 \qquad (2.4)$$

Comparing the coefficients of like terms of x, on either side, we obtain the relation

$$(n+1)a_{n+1} - 2a_{n-1} = 0 \quad \text{for} \quad n = 1,2,3 \ldots$$

and this is a recurrence relation leading to the determination of a_n's.

Put $n = 1$ in the recurrence relation to obtain $a_2 = a_0$.

Similarly, $n = 2$ gives

$$a_3 = \frac{2}{3}a_1$$

and $n = 3$ gives

$$a_4 = \frac{1}{2}a_2 = \frac{1}{2}a_0$$

and so on.

We observe that the coefficients are in terms of a_0 and a_1. As the given equation is a first-order equation, its general solution involves only one arbitrary constant. Hence, we need to find out whether there is a relation between and a_1 (if possible), or sometimes we obtain $a_1 = 0$.

For this, collect the coefficient of x^1 from Equation (2.3); we obtain that $a_1 = 0$ and thus

$$a_3 = a_5 = \ldots = 0$$

Hence, the power series solution is

$$y(x) = a_0\left(1 + x^2 + \frac{1}{2}x^4 + \ldots\right)$$

with $a_0 = 1$ in view of the given initial condition $y(0) = 1$.

Further, we can conclude that this series solution is valid for $|x| < \infty$, as can be seen in the plot in Figure 2.1.

We now consider a variable coefficient equation.

Example 2.12

Solve

$$(x-1)y' + y = 0$$

with $y(4) = 5$.

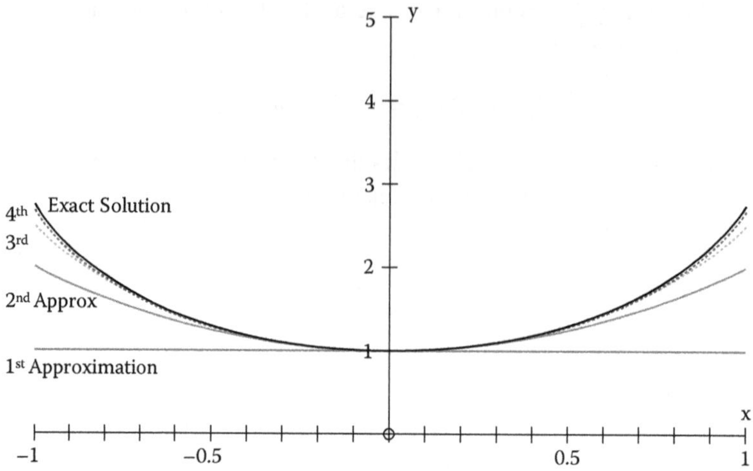

Figure 2.1 Comparison between exact solution (black) and truncated solutions.

Solution: This equation can be expressed as

$$y' + \frac{y}{(x-1)} = 0.$$

Because $P(x) = \frac{1}{x-1}$, the region of analyticity of the power series of $P(x)$ about $x = 4$ is $|x-4| < 3$. Here, x_0 is given to be 4, which is an ordinary point of the given differential equation. Taking $y = \sum_{n=0}^{\infty} a_n (x-4)^n$ and substituting in the given equation, we have

$$(x-4)\sum_{n=1}^{\infty} na_n (x-4)^{n-1} + 3\sum_{n=1}^{\infty} na_n (x-4)^{n-1} + \sum_{n=0}^{\infty} a_n (x-4)^n = 0 \tag{2.5}$$

We notice that the a_n's are governed by the recurrence relation

$$a_{n+1} = -\frac{1}{3} a_n$$

for $n = 0,1,2 \ldots$.

This gives

$$a_1 = -\frac{1}{3} a_0$$

$$a_2 = -\frac{1}{3} a_1 = \frac{1}{9} a_0$$

and so on.

Thus, the power series solution is

$$y(x) = a_0\left(1 - \frac{1}{3}(x-4) + \frac{1}{9}(x-4)^2 + \ldots\right)$$

with $a_0 = 5$, in view of the given initial condition.

Note: The exact solution to this IVP is

$$y(x) = \frac{15}{x-1}$$

Figure 2.2 shows the plot of the exact solution along with the first six truncated solutions. We can see that as we increase the number of terms in the series solution, the solution becomes closer and closer to the exact one. Also, as mentioned in the solution, the power series solution is valid only on the interval (1,7), beyond which the solution diverges from the exact solution.

Now, let us consider a nonhomogeneous first-order equation.

Example 2.13

Find the solution of the nonhomogeneous equation $y' + 3y = 8$ with $y(0) = 4$.

Solution: Here, x_0 is given to be 0, which is an ordinary point of the given differential equation.

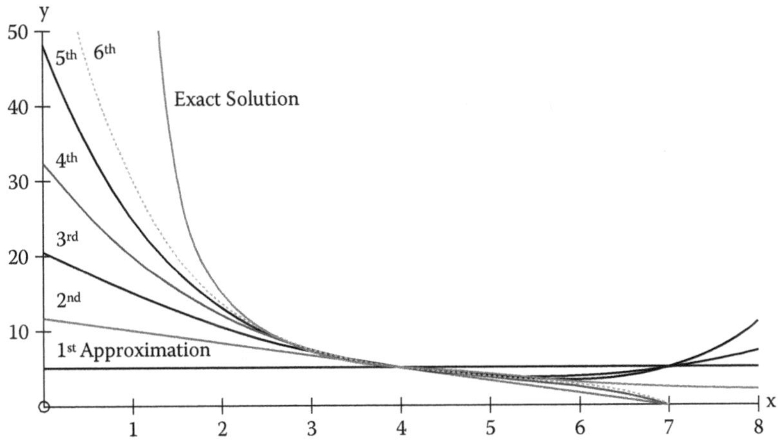

Figure 2.2 Comparison between exact solution and the first six truncated solutions.

Taking $y = \sum_{n=0}^{\infty} a_n x^n$ and substituting in the given equation, we have

$$\sum_{n=1}^{\infty} n a_n x^{n-1} + 3\sum_{n=0}^{\infty} a_n x^n = 8 \tag{2.6}$$

Comparing the coefficient of x^0: $a_1 = 8 - 3a_0$.

Comparing the coefficient of x^n, the recurrence relation is $a_{n+1} = -\frac{3}{n+1} a_n$ for $n = 1,2 \ldots$.

This gives

$$a_2 = -\frac{3}{2}(8 - 3a_0)$$

$$a_3 = \frac{3}{2}(8 - 3a_0)$$

and so on.

Thus, the power series solution is

$$y(x) = a_0\left(1 - 3x + \frac{9}{2}x^2 + \ldots\right) + 8x\left(1 - \frac{3}{2}x + \frac{3}{2}x^2 + \ldots\right) \tag{2.7}$$

with $a_0 = 4$ and is valid for $|x| < \infty$.

Note that in the power series solution Equation (2.7), the first series represents the general solution of the given equation, whereas the second part is the particular solution corresponding to the nonhomogeneous right-hand-side part.

Note: The exact solution is

$$y(x) = \frac{4}{3}\left(e^{-3x} + 2\right).$$

Figure 2.3 depicts the exact and truncated series solutions to this IVP.

Example 2.14

Solve the nonlinear IVP $y' = 1 + y^2$ with $y(0) = 0$.

Solution: Here, x_0 is given to be 0, which is an ordinary point of the given differential equation. Choosing $y = \sum_{n=0}^{\infty} a_n x^n$, we have

$$\sum_{n=1}^{\infty} n a_n x^{n-1} = \left(1 + \sum_{n=0}^{\infty} a_n x^n\right)^2 \tag{2.8}$$

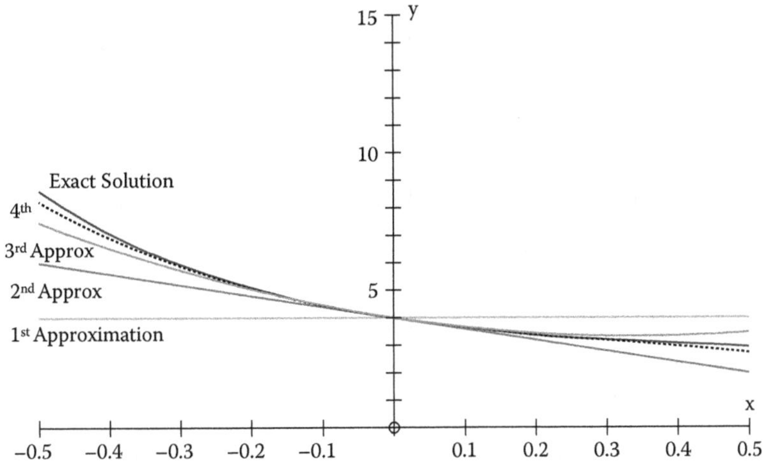

Figure 2.3 Comparison between exact solution and truncated solutions to nonhomogeneous equation $y' + 3y = 8$.

Comparing the coefficient of different powers of x, we obtain

$$a_1 = 1 + a_0^2$$

As $a_0 = 0$ because of the given initial condition, we have $a_1 = 1$.
Further, $a_2 = a_0 a_1$; thus,

$$a_2 = 0$$

and

$$3a_3 = a_1^2 + 2a_0 a_2 + 2a_2$$

which gives

$$a_3 = \frac{1}{3}\dots$$

Thus, the power series solution up to the third power of x is given by

$$y(x) = x + \frac{x^3}{3} + \frac{2x^5}{15} + \dots$$

The nonlinearity and the nonhomogeneous nature of the given equation complicates the computation of the successive coefficients in the power series solution. Figure 2.4 shows the comparison of approximate solutions with the exact solution.

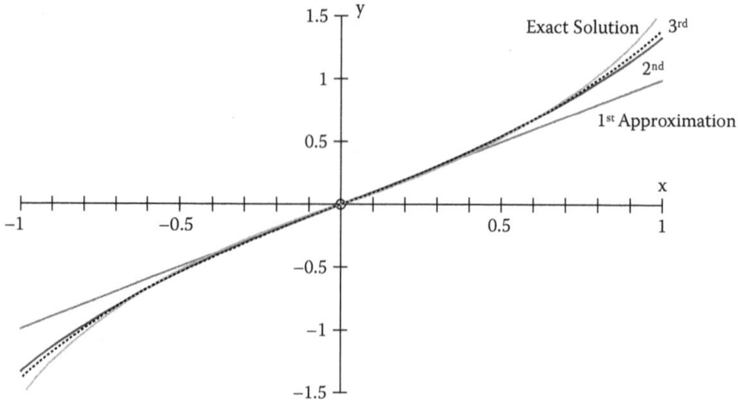

Figure 2.4 Comparison of approximate solutions with the exact solutions.

Note: The exact solution is $y(x) = \tan x$.

Now, we find series solutions to higher-order equations.

Example 2.15

Find the solution near $x = 1$ of $y'' + (x-1)y = e^x$.

Solution: Here, x_0 is given to be 1, which is an ordinary point of the given differential equation. Taking $y = \sum_{n=0}^{\infty} a_n (x-1)^n$, from the given equation, we have

$$\sum_{n=2}^{\infty} n(n-1)a_n (x-1)^{n-2} + \sum_{n=0}^{\infty} a_n (x-1)^{n+1} = e^x \tag{2.9}$$

Notice that

$$e^x = e^{(x-1)+1} = e\left\{1 + (x-1) + \frac{(x-1)^2}{2!} + \ldots\right\}$$

After a straightforward comparison of the coefficients of like powers of x, we have the recurrence relation

$$a_{n+2} = \frac{\dfrac{e}{n!} - a_{n-1}}{(n+1)(n+2)}$$

for $n = 1,2,3 \ldots$.

Further, $a_2 = \frac{e}{2}$; now, plugging in the values of n in the recurrence relation, we have

$$a_3 = \frac{\frac{e}{1!} - a_0}{2.3}$$

$$a_4 = \frac{\frac{e}{2!} - a_1}{3.4}$$

and so on.

Thus, the power series solution is

$$y(x) = a_0\left(1 + \frac{(x-1)^3}{6} - \ldots\right) + a_1\left((x-1) - \frac{(x-1)^4}{12} + \ldots\right)$$
$$+ e\left(\frac{(x-1)^2}{2} - \frac{(x-1)^3}{6} + \ldots\right),$$

which is valid for $|x| < \infty$. Here, a_0 and a_1 are two arbitrary constants in the solution. This is to be expected as we have considered a second-order differential equation.

Example 2.16

Solve the nonlinear second-order IVP $y'' + y^2 = 0$ with $y(0) = 1$, $y'(0) = 0$.

Solution: Here, x_0 is given to be 0, which is an ordinary point of the given differential equation.

Taking $y = \sum_{n=0}^{\infty} a_n x^n$, we have

$$\sum_{n=2}^{\infty} n(n-1) a_n x^{n-2} + \left(\sum_{n=0}^{\infty} a_n x^n\right)^2 = 0 \qquad (2.10)$$

Further, as $y(0) = 1$, we obtain $a_0 = 1$.

Comparing the coefficient of like powers of x

$$a_2 = -\frac{a_0^2}{2} = -\frac{1}{2}$$

Further, $6a_3 + 2a_0a_1 = 0$; thus, $a_3 = 0$.

$$12a_4 + a_1^2 + 2a_0a_2 = 0$$

which gives

$$a_4 = \frac{1}{12}\ldots$$

Thus, the power series solution is

$$y(x) = 1 - \frac{x^2}{2} + \frac{x^4}{12} + \ldots.$$

In view of the nonlinearity of the differential equation, the calculation of the terms involving higher-powers of x can be prohibitively laborious.

Example 2.17

Find the solution near $x = 0$ for $y''' - y = 0$.

Solution: Here, x_0 is given to be 0, which is an ordinary point of the given differential equation.

Taking

$$y = \sum_{n=0}^{\infty} a_n x^n,$$

we obtain

$$\sum_{n=3}^{\infty} n(n-1)(n-2)a_n x^{n-3} - \sum_{n=0}^{\infty} a_n x^n = 0$$

The recurrence relation leading to the determination of a_n's is

$$a_{n+3} = \frac{a_n}{(n+1)(n+2)(n+3)} \quad \text{for} \quad n = 0,1,2 \ldots.$$

Plugging the values of n into the recurrence relation, we have

$$a_3 = \frac{a_0}{1.2.3}$$

$$a_4 = \frac{a_1}{2.3.4}$$

$$a_5 = \frac{a_2}{3.4.5}$$

and so on.

Thus, the power series solution is

$$y(x)=a_0\left(1+\frac{x^3}{6}+\ldots\right)+a_1\left(x+\frac{x^4}{24}+\ldots\right)+a_2\left(x^2+\frac{x^5}{60}+\ldots\right),$$

which is valid for $|x|<\infty$.

Notice that we have the arbitrary constants a_0, a_1, a_2 in the general solution as we started with a third-order differential equation.

We now discuss the method for finding solutions to the differential equations at the points called singular points.

Solution at a Singular Point (Regular) of an Ordinary Differential Equation

Definition

Recall from the previous section that x_0 is a singular point of

$$y'+P(x)y=0, \quad y(x_0)=y_0 \tag{2.11}$$

if $P(x)$ is not analytic at x_0. We say that x_0 is a regular singular point (RSP) if the singularity of $P(x)$ is not worse than $\frac{1}{x-x_0}$, that is, the expansion of $P(x)$ about $x=x_0$ is of the form

$$P(x)=\frac{b_1}{x-x_0}+\sum_{n=0}^{\infty}a_n\left(x-x_0\right)^n$$

where a_n's and b_1 are constants.

Otherwise, x_0 is termed an IRSP.

For instance,

(i) $x=0$ is an RSP of

$$y'-\frac{\cos x}{x}y=0.$$

Note that

$$P(x)=-\frac{\cos x}{x}=-\frac{1}{x}+\frac{x}{2}-\frac{x^3}{24}+\ldots$$

(ii) $x=1$ is an IRSP of

$$y' - \frac{1}{(x-1)^2}\, y = 0.$$

Here,

$$P(x) = -\frac{1}{(x-1)^2}.$$

From these examples, we observe that x_0 is an RSP of Equation (2.11) if $(x-x_0)P(x)$ is analytic at x_0. ■

At RSPs, we look for solutions of the form

$$y = \sum_{n=0}^{\infty} a_n (x-x_0)^{n+m}, \quad a_0 \neq 0$$

because if $a_0 = 0$, then some positive integral power of x can be factored out of the power series part and can be combined with $(x-x_0)^n$. This series is called the Frobenius series, and the method is called the Frobenius method.

We now consider a second-order differential equation and define the conditions for a point x_0 to be called an RSP.

Definition

In particular, we say that x_0 is an RSP of the second-order equation

$$y'' + P(x)y' + Q(x)y = 0$$

if

$$(x-x_0)P(x)$$

and

$$(x-x_0)^2 Q(x)$$

are both analytic at x_0.

For example, for the equation

$$y'' + \frac{(\sin x)}{x} y = 0,$$

$x = 0$ is an RSP, whereas for

$$y'' + \frac{(\sin x)}{x^3} y = 0,$$

$x = 0$ is an IRSP. (You may easily verify these results.) ■

Frobenius Series Method

Let us see the Frobenius method for finding a series solution to an ODE at a point that is an RSP of it. The method is described for a second-order linear equation. It can as well be applied for higher-order equations.

If x_0 is an RSP of

$$y'' + P(x)y' + Q(x)y = 0 \tag{2.12}$$

assume a series solution of the form

$$y = \sum_{n=0}^{\infty} a_n (x - x_0)^{n+m}$$

where $a_0 \neq 0$.

Then, find

$$y' = \sum_{n=0}^{\infty} (n+m) a_n (x - x_0)^{n+m-1}$$

$$y'' = \sum_{n=0}^{\infty} (n+m)(n+m-1) a_n (x - x_0)^{n+m-2}$$

and substitute in Equation (2.12).

Collect the coefficient of least power of x that is going to be an equation involving m. This equation is called the indicial equation. For the second-order equation as in Equation (2.12), the indicial equation is a second-degree equation in m that, on solving, gives two roots, say, m_1 and m_2.

Now, three cases arise:

(i) Roots differ by an integer, in which case Equation (2.12) has two Frobenius solutions, $y_1(x)$ and $y_2(x)$, that are linearly independent ($y_1(x) \neq k y_2(x)$, for any scalar k). And, the solutions are

$$y_1 = x^{m_1}\left(\sum_{n=0}^{\infty} a_n (x - x_0)^n\right)_{a_n\text{'s evaluated at } m=m_1}$$

$$y_2 = x^{m_2}\left(\sum_{n=0}^{\infty} a_n (x - x_0)^n\right)_{a_n\text{'s evaluated at } m=m_2}$$

(ii) Roots differ by an integer. Let $m_1 - m_2 = n$ (a positive integer). Then, one Frobenius series solution is given by

$$y_1 = x^{m_1}\left(\sum_{n=0}^{\infty} a_n (x - x_0)^n\right)_{a_n\text{'s evaluated at } m=m_1}$$

and the second solution y_2 is obtained using the formula

$$y_2 = y_1\left(\int\left(\frac{1}{y_1^2}e^{-\int P(x)dx}\right)dx\right).$$

It may be noted that we may sometimes obtain terms like $\log(x)$ in y_2. In such a case, the solution cannot be a Frobenius series solution. Thus, we have that, if the two roots of the indicial equation differ by an integer, one Frobenius series solution always exists, whereas the existence of a second Frobenius series solution depends on the problem considered. This is shown in the examples considered in the following material.

(iii) Roots are equal (each equal to m), in which case there is only one Frobenius series solution, which is given by

$$y_1 = x^m \left(\sum_{n=0}^{\infty} a_n (x - x_0)^n \right)_{a_n\text{'s evaluated at } m=m}$$

The second solution can be obtained using the same formula for y_2 as in case (ii).

Here, the second solution looks like

$$y_2 = y_1 \log x + \sum_{n=0}^{\infty} b_n (x - x_0)^n$$

where b_n's are found by using the fact that y_2 is also a solution to the given differential equation.

Example 2.18

To illustrate case (i), consider the following problem:

Solve in series, the equation

$$5x(1-x)y'' - y' + 4y = 0$$

about $x = 0$.

Solution: Here, $x_0 = 0$ is a RSP of the given differential equation (from the definition of an RSP).

Taking $y = \sum_{n=0}^{\infty} a_n x^{n+m}$, where $a_0 \neq 0$, and substituting in the given differential equation, we have

$$5\sum_{n=0}^{\infty}(n+m)(n+m-1)a_n x^{n+m-1} - 5\sum_{n=0}^{\infty}(n+m)(n+m-1)a_n x^{n+m}$$

$$-\sum_{n=0}^{\infty}(n+m)a_n x^{n+m-1} + 4\sum_{n=0}^{\infty} a_n x^{n+m} = 0 \qquad (2.13)$$

Collecting the coefficient of the least power of x, that is, x^{m-1} from Equation (2.13), we obtain the indicial equation as

$$(5m(m-1) - m)a_0 = 0.$$

Solving this equation for its roots, we obtain

$$m=0, \quad \frac{6}{5}.$$

Now, collecting the coefficient of x^{m+n} from Equation (2.13), we obtain the recurrence relation as

$$a_{n+1}=-\frac{5(m+n)(m+n-1)-4}{(m+n+1)(5m+5n-1)}a_n, \quad n=0,1,2\ldots \quad (2.14)$$

Because the roots of the indicial equation do not differ by an integer, we can find the two linearly independent solutions using the formulae mentioned for case (i):

For $m = 0$, the coefficients can be obtained from Equation (2.14) as follows:

$$a_1=-\frac{1}{2}a_0$$

$$a_2=-\frac{1}{9}a_0$$

$$a_3=-\frac{13}{252}a_0$$

and so on.

Thus, one solution of the given problem is

$$y_1(x)=a_0\left\{1-\frac{x}{2}-\frac{x^2}{9}-\ldots\right\}.$$

For $m=\frac{6}{5}$, again using Equation (2.14), we have

$$a_1=\frac{23}{80}a_0$$

$$a_2=\frac{299}{2100}a_0$$

and so on.

Thus, the second solution is

$$y_2(x)=a_0x^{6/5}\left\{1+\frac{23}{80}x+\frac{299}{2100}x^2+\ldots\right\}.$$

Now, the general solution to the given equation is

$$y(x)=c_1y_1(x)+c_2y_2(x)$$

where c_1 and c_2 are arbitrary constants.

Example 2.19

To illustrate case (ii), consider the problem of finding two linearly independent solutions of

$$x^2y'' + xy' + \left(x^2 - \frac{1}{4}\right)y = 0$$

about $x = 0$.

Solution: Here, $x_0 = 0$ is an RSP of the given differential equation.

Substituting $y = \sum_{n=0}^{\infty} a_n x^{n+m}$, where $a_0 \neq 0$, in the given equation, we have

$$\sum_{n=0}^{\infty}(n+m)(n+m-1)a_n x^{n+m} - \sum_{n=0}^{\infty}(n+m)a_n x^{n+m} + \sum_{n=0}^{\infty} a_n x^{n+m+2}$$
$$-\frac{1}{4}\sum_{n=0}^{\infty} a_n x^{n+m} = 0 \qquad (2.15)$$

The indicial equation is

$$\left(m^2 - \frac{1}{4}\right)a_0 = 0$$

which gives

$$m = -\frac{1}{2}, \ \frac{1}{2}$$

Observe that the two roots differ by an integer.

Here, $n = 1$ and for $m_1 = \frac{1}{2}$, the recurrence relation is

$$a_n = -\frac{a_{n-2}}{n(n+1)}, \quad n = 2, 3, 4\ldots$$

We see from this relation that $a_2, a_4, a_6\ldots$ are in terms of a_0, whereas $a_3, a_5, a_7\ldots$ are in terms of a_1. We know that $a_0 \neq 0$. Let us find the value of a_1. For this, collect the coefficient of x^{m+1} for Equation (2.15), which gives $a_1 = 0$.

Now, using the recurrence relation given previously in this example, we obtain

$$a_2 = -\frac{a_0}{3!}$$

$$a_3 = -\frac{a_1}{4.3} = 0$$

$$a_4 = -\frac{a_2}{5.4} = \frac{a_0}{5!}$$

and so on.

Thus, the first Frobenius solution is

$$y_1(x) = a_0 x^{1/2}\left\{1 - \frac{x^2}{2!} + \frac{x^4}{5!} - \ldots\right\}.$$

Now,

$$y_2 = a_0^* x^{1/2}\left\{1 - \frac{x^2}{2!} + \frac{x^4}{5!} - \ldots\right\}\left(\int \frac{1}{\left(x^{1/2}\left\{1 - \frac{x^2}{2!} + \frac{x^4}{5!} - \ldots\right\}\right)^2} e^{-\int \frac{1}{x} dx} dx\right)$$

After straightforward but lengthy calculations, we obtain

$$y_2 = a_0^* x^{-1/2}\left\{1 - \frac{x^2}{2!} + \frac{x^4}{4!} - \ldots\right\}$$

which is the second Frobenius solution to the given equation. Thus, this problem has two Frobenius series solutions.

Example 2.20

To illustrate case (iii), consider the following problem: Solve

$$x^2 y'' + xy' + x^2 y = 0$$

using the Frobenius method.

Solution: Here, x_0 is taken to be 0, which is an RSP of the given differential equation. If $y = \sum_{n=0}^{\infty} a_n x^{n+m}$, where $a_0 \neq 0$, then we have

$$\sum_{n=0}^{\infty}(n+m)(n+m-1)a_n x^{n+m} - \sum_{n=0}^{\infty}(n+m)a_n x^{n+m} + \sum_{n=0}^{\infty} a_n x^{n+m+2} = 0$$

The indicial equation is $m^2 a_0 = 0$, which gives $m = 0,0$.

Hence, only one Frobenius series solution exists.

The recurrence relation for $m = 0$ is

$$a_n = -\frac{a_{n-2}}{n^2}, \quad n = 2,3,4\ldots.$$

Following the same argument as in the previous example, we obtain that $a_1 = 0$.

Thus,

$$a_2 = -\frac{a_0}{2^2}$$

$$a_3 = -\frac{a_1}{3^2} = 0$$

$$a_4 = -\frac{a_2}{4^2} = \frac{a_0}{2^2 4^2}$$

and so on.

Thus,

$$y_1(x) = a_0\left\{1 - \frac{x^2}{2^2} + \frac{x^4}{2^2.4^2} - \ldots\right\}.$$

Let

$$y_2 = y_1 \log x + \sum_{n=0}^{\infty} b_n x^n$$

be the second solution to the given equation; then,

$$y_2' = y_1' \log x + \frac{y_1}{x} + \sum_{n=1}^{\infty} n b_n x^{n-1}$$

$$y_2'' = y_1'' \log x + \frac{2y_1'}{x} - \frac{y_1}{x^2} + \sum_{n=2}^{\infty} n(n-1) b_n x^{n-2}$$

and substituting y_2, y_2', and y_2'' in the given equation, we have

$$\log x(x^2 y_1'' + x y_1' + x^2 y_1) + 2x y_1' - y_1 + \sum_{n=2}^{\infty} n(n-1) b_n x^n$$

$$+ \sum_{n=1}^{\infty} n b_n x^n + \sum_{n=0}^{\infty} b_n x^{n+2} = 0$$

As y_1 is a solution of the equation, the first term becomes zero. Substituting the expression for y_1 and comparing the coefficients of different powers of x, we have

$$b_1 = 0, b_3 = 0\ldots$$

$$b_2 = \frac{1}{2^2} - \frac{b_0}{2^2}$$

$$b_4 = -\frac{1+\frac{1}{2}}{2^2.4^2} + \frac{b_0}{2^2.4^2}$$

$$b_6 = -\frac{1+\frac{1}{2}+\frac{1}{3}}{2^2.4^2.6^2} + \frac{b_0}{2^2.4^2.6^2}$$

Thus,

$$y_2(x) = y_1 \log x + b_0 y_1 + \frac{x^2}{2^2} - \frac{1+\frac{1}{2}}{2^2.4^2} x^4 + \ldots$$

and hence the general solution for the given equation can be found.

We can apply the Frobenius method to solve a first-order differential equation at its RSP as shown in the next example.

Example 2.21

Solve

$$y' - \frac{y}{x} = 0$$

using the Frobenius method.

Solution: Here, x_0 is taken to be 0, which is an RSP of the given differential equation.

Let

$$y = \sum_{n=0}^{\infty} a_n x^{n+m}$$

where $a_0 \neq 0$, then substituting y, y' in the given equation, we obtain

$$\sum_{n=0}^{\infty} (n+m) a_n x^{n+m} - \sum_{n=0}^{\infty} a_n x^{n+m} = 0$$

The indicial equation is

$$(m-1)a_0 = 0$$

which gives $m = 1$.

The recurrence relation is

$$a_n = 0$$

for $n = 1, 2, 3 \ldots.$

Thus, the Frobenius series solution is

$$y(x) = a_0 x$$

which is valid for $|x| < \infty$ [the region of analyticity of $xP(x)$].

Example 2.22

Now, we examine an example in which the two roots of the indicial equation differ by an integer, but unlike in Example 2.19, we show that it has no second Frobenius solution.

Solve

$$x^2 y'' + xy' + (x^2 - 1)y = 0$$

using the Frobenius method.

Solution: Here, x_0 is taken to be 0, which is an RSP of the given differential equation.

Let

$$y = \sum_{n=0}^{\infty} a_n x^{n+m}$$

where $a_0 \neq 0$, substituting it in the given equation, we obtain

$$\sum_{n=0}^{\infty}(n+m)(n+m-1)a_n x^{n+m} + \sum_{n=0}^{\infty}(n+m)a_n x^{n+m} + \sum_{n=0}^{\infty} a_n x^{n+m+2} - \sum_{n=0}^{\infty} a_n x^{n+m} = 0 \tag{2.16}$$

The indicial equation is

$$(m^2 - 1)a_0 = 0$$

which gives $m = -1, 1$.

Thus, the two roots differ by an integer.

Here, $n = 2$ and for $m_1 = 1$, the recurrence relation is

$$a_n = -\frac{a_{n-2}}{n(n+2)}, \quad n = 2,3,4\ldots$$

Hence, we see from this relation that $a_2, a_4, a_6 \ldots$ are in terms of a_0, whereas $a_3, a_5, a_7 \ldots$ are in terms of a_1. We know that $a_0 \neq 0$. Let us find the value of a_1. For this, collect the coefficient of x^{m+1} for Equation (2.16), which gives $a_1 = 0$.

Now, using the recurrence relation, we obtain

$$a_2 = -\frac{a_0}{2^2 2!}$$

$$a_3 = -\frac{a_1}{15} = 0$$

$$a_4 = -\frac{a_2}{4.6} = \frac{a_0}{2^4 2!3!}$$

and so on.

Thus, the first Frobenius solution is

$$y_1(x) = a_0 x\left\{1 - \frac{x^2}{2^2 2!} + \frac{x^4}{2^4 2!3!} - \ldots\right\}.$$

Let us find the second solution using the formula

$$y_2 = y_1\left(\int\left(\frac{1}{y_1^2} e^{-\int P(x)dx}\right)dx\right).$$

Hence,

$$y_2 = \left(x\left\{1 - \frac{x^2}{2^2 2!} + \frac{x^4}{2^4 2!3!} - \ldots\right\}\right)\left(\int\left(\frac{1}{\left(x\left\{1 - \frac{x^2}{2^2 2!} + \frac{x^4}{2^4 2!3!} - \ldots\right\}\right)^2} e^{-\int \frac{1}{x}dx}\right)dx\right)$$

$$= x\left\{1 - \frac{x^2}{2^2 2!} + \frac{x^4}{2^4 2!3!} - \ldots\right\}\left\{\frac{\log x}{4} - \frac{1}{2x^2} + \frac{5x^2}{384} - \ldots\right\}$$

We can observe that this solution has the logarithmic function and thus cannot form a Frobenius series solution. Hence, in this case, we have only one Frobenius series solution to the problem.

Remarks on the Frobenius Solution at Irregular Singular Points

Let us now discuss the existence of the Frobenius solution at the IRSPs of a differential equation. For this, first consider the ODE

$$y' = \frac{y}{x^2}.$$

Clearly, $x = 0$ is an IRSP. If we try taking its solution as

$$y = \sum_{n=0}^{\infty} a_n x^{n+m}, \quad a_0 \neq 0,$$

then we have

$$\sum_{n=0}^{\infty} (n+m) a_n x^{n+m+2} - \sum_{n=0}^{\infty} a_n x^{n+m} = 0$$

and the indicial equation is $a_0 = 0$, which is not possible.

Hence, there is no Frobenius solution to this problem.

Now, consider, another equation:

$$y' + \frac{y}{x-4} = 0, \quad y(4) = 1.$$

Again, taking a Frobenius series solution about $x = 4$ as

$$y = \sum_{n=0}^{\infty} a_n (x-4)^{n+m}, \quad a_0 \neq 0$$

we have

$$\sum_{n=0}^{\infty} (n+m) a_n (x-4)^{n+m} - \sum_{n=0}^{\infty} a_n (x-4)^{n+m} = 0$$

and the indicial equation is

$$(m+1) a_0 = 0$$

which gives $m = -1$.

The recurrence relation is

$$(n+m+1) a_n = 0.$$

Hence, $a_n = 0$ for $n = 1, 2, 3 \ldots.$

Thus, the solution is

$$y = \frac{a_0}{x-4}$$

which is not defined at $x = 4$.

Now, consider the equation

$$x^2 y'' + (3x-1)y' + y = 0$$

so clearly, $x = 0$ is an IRSP.

Assuming the Frobenius series solution to it gives

$$\sum_{n=0}^{\infty}(n+m)(n+m-1)a_n x^{n+m} + 3\sum_{n=0}^{\infty}(n+m)a_n x^{n+m}$$
$$-\sum_{n=0}^{\infty}(n+m)a_n x^{n+m-1} + \sum_{n=0}^{\infty}a_n x^{n+m} = 0$$

Thus, the indicial equation is $m = 0$.

The recurrence relation

$$a_{n+1} = (n+m+1)a_n, \quad n = 0,1,2\ldots.$$

For $m = 0$, the solution is

$$y = \sum_{n=0}^{\infty} n!x^n$$

which converges only at $x = 0$.

We conclude by saying that, at IRSPs, the Frobenius method fails to provide valid solutions to the problems. In fact, there is no method for finding solutions at IRSPs (Paris and Wood, 1986) that are finite. But, at IRSPs that are at infinity, we have the asymptotic method, which is explained in Chapter 3.

Taylor Series Method

Some of the differential equations of the form $y' = f(x, y)$ with the condition $y(x_0) = y_0$ can be solved by another method called the Taylor series method. We state here some theorems from Tenenbaum and Pollard (1985) that are useful for the method.

Theorem 2.1

Consider the differential equation $y' = f(x, y)$. If the function $f(x, y)$ is analytic at a point (x_0, y_0), that is, it has a Taylor series expansion in powers of $(x - x_0)$ and $(y - y_0)$ that is valid in some rectangle $|x - x_0| < r$, $|y - y_0| < r$, which has (x_0, y_0) as its center. If for every (x, y) in this $2r \times 2r$ rectangle, $|f(x, y)| \leq M$ where M is a positive number, then there is a unique particular solution $y(x)$ that is analytic at $x = x_0$, satisfying the condition $y(x_0) = y_0$, which is valid in an interval $|x - x_0| < \min\left\{r, \frac{r}{3M}\right\}$. ■

Theorem 2.2

If $f(x)$ is infinitely derivable, then the Taylor series of $f(x)$ about $x = x_0$ is

$$f(x) = \sum_{n=0}^{\infty} \frac{f^n(x_0)}{n!}(x - x_0)^n.$$

■

Theorem 2.3

The Taylor expansion for a function of two variables x, y, say $f(x, y)$ about the point (x_0, y_0), is

$$f(x, y) = f(x_0, y_0) + \left\{(x - x_0)\left(\frac{\partial f}{\partial x}\right)_{(x_0, y_0)} + (y - y_0)\left(\frac{\partial f}{\partial y}\right)_{(x_0, y_0)}\right\} +$$

$$\frac{1}{2!}\left\{(x - x_0)^2\left(\frac{\partial^2 f}{\partial x^2}\right)_{(x_0, y_0)} + (y - y_0)\left(\frac{\partial^2 f}{\partial y^2}\right)_{(x_0, y_0)}\right.$$

$$\left. + 2(x - x_0)(y - y_0)\left(\frac{\partial^2 f}{\partial x \partial y}\right)_{(x_0, y_0)}\right) + \ldots$$

■

In this section, we explain the Taylor series method to find the series solution of a differential equation through some examples. This method, also called the method of successive differentiation, also gives the region of convergence of the power series solution for nonlinear equations.

Example 2.23

Find a series solution of $y' = y + x$ with $y(0) = 1$ using the Taylor series.

Solution: Given that

$$y' = y + x \tag{2.17}$$

Using $y(0) = 1$, we obtain $y'(0) = 1$.

Now, differentiating Equation (2.17) with respect to x, we obtain

$$y'' = y' + 1 \tag{2.18}$$

and thus $y''(0) = 1$.

Again, differentiating Equation (2.18) with respect to x, we obtain $y''' = y''$, and hence $y'''(0) = 1$ and so on.

Thus, from Theorem 2.2, we form the solution to the given IVP as

$$y(x) = 1 + x + x^2 + \frac{x^3}{3} + \dots.$$

Figure 2.5 shows the plot of exact and approximate solutions up to the first four terms of the problem presented.

Example 2.24

Find a series solution for the equation $y' = y^2$ with $y(0) = 1$ using the Taylor series.

Solution: Given

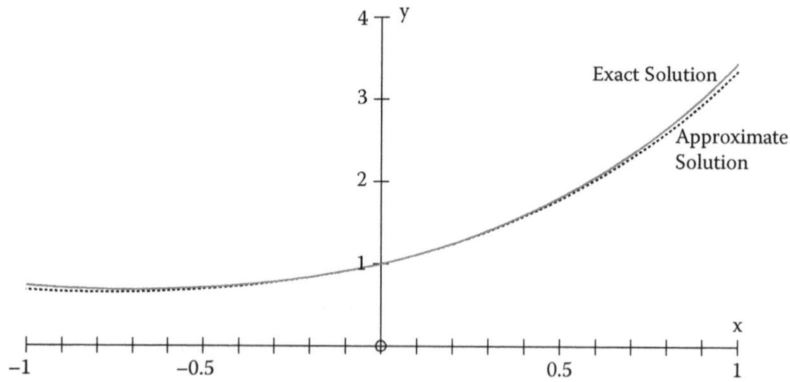

Figure 2.5 Plot of exact and approximate solutions of Equation (2.17).

$$y' = y^2 \tag{2.19}$$

Using the given condition, we obtain

$$y'(0) = 1.$$

Differentiating Equation (2.19) with respect to x, we have

$$y'' = 2yy'. \tag{2.20}$$

Thus, $y''(0) = 2.$

Also,

$$y''' = 2((y')^2 + yy'')$$

and

$$y'''(0) = 6\dots.$$

Thus, the solution to the given IVP is

$$y(x) = 1 + x + x^2 + x^3 + \dots.$$

You may note that the subsequent terms of the series can be found in principle.

The exact solution to this IVP is

$$y(x) = \frac{1}{(1-x)}.$$

Figure 2.6 shows the plots of the exact and the approximate solution up to the first four terms of the problem. The reason for the deviation of the approximate solution from the exact solution is that the series solutions have a certain region of convergence where it coincides with the exact solution. Example 2.27 calculates the region of convergence of the solution to this problem.

Example 2.25

Find a series solution of $y'' + y' - xy = 0$ with $y(0) = 1$ $y'(0) = 0$ using the Taylor series.

Solution: Given that

$$y'' = -y' + xy \tag{2.21}$$

Using the given initial conditions, we obtain $y''(0) = 0.$

Differentiating Equation (2.21), once with respect to x, we obtain

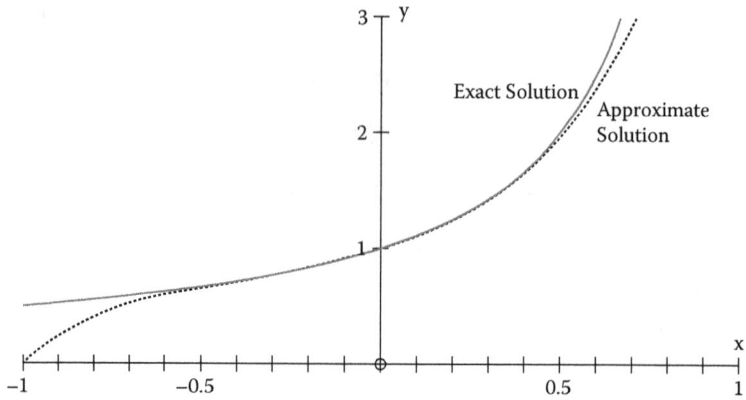

Figure 2.6 Plot of exact and approximate solutions of Equation (2.19).

$$y''' = -y'' + xy' + y \tag{2.22}$$

and hence $y'''(0) = 1$,

Again, differentiating Equation (2.22) with respect to x, we obtain

$$y^{IV} = -y''' + xy'' + 2y'$$

and

$$y^{IV}(0) = -1\ldots.$$

Now, substituting these values in the Taylor expansion for $y(x)$, we obtain that

$$y(x) = 1 + \frac{x^3}{3!} - \frac{x^4}{4!}\ldots.$$

Example 2.26

Find a series solution of the Hermite equation

$$y'' - 2xy' + 2\lambda y = 0$$

where λ is a constant.

Solution: Because initial conditions are not provided in the problem, let us assume the conditions

$$y(0) = a_0, \quad y'(0) = a_1$$

From the given equation, we have

$$y'' = 2xy' - 2\lambda y$$

Thus,

$$y''(0) = -2\lambda a_0$$

Now,

$$y''' = 2xy'' + 2y' - 2\lambda y'$$

and

$$y'''(0) = 2a_1 - 2\lambda a_1$$

Also,

$$y^{IV} = 2xy''' + 4y'' - 2\lambda y''$$
$$y^{IV}(0) = -4\lambda a_0 + 4\lambda^2 a_0 \ldots$$

Thus, the solution to the given equation is

$$y(x) = a_0\left(1 - \frac{2\lambda}{2!}x^2 + \frac{2^2\lambda(\lambda-2)}{4!}x^4 + \ldots\right) + a_1\left(x - \frac{2(\lambda-1)}{3!}x^3 + \ldots\right)$$

where a_0 and a_1 are two arbitrary constants in the solution chosen. If the conditions on y and y' are given at $x = 0$, then a_0 and a_1 can be determined.

Example 2.27

Find the interval of convergence of the Taylor series solution of $y' = y^2$ with $y(0) = 1$.

Solution: Here,

$$f(x, y) = y^2, \quad (x_0, y_0) = (0,1).$$

As mentioned in Theorem 2.1, let us choose the rectangle $|x| < r$, $|y-1| < r$ with center at $(0,1)$.

Then,

$$|f(x, y)| = y^2 = 1 + 2(y-1) + (y-1)^2$$

(from Theorem 2.3).

Thus,

$$\left|f(x,y)\right| < 1 + 2r + r^2.$$

Hence, by Theorem 2.1, the Taylor series solution is convergent in the interval

$$I : |x| < \min\left\{r, \frac{r}{3(r+1)^2}\right\}.$$

Let

$$g(r) = \frac{r}{3(r+1)^2}.$$

Using the theory of maxima and minima, we can show that this function h is maximum at $r = 1$.

Hence, the interval of convergence is

$$I : |x| < \min\left\{1, \frac{1}{12}\right\}$$

which is

$$|x| < \frac{1}{12}.$$

Theorem 2.4: Extension of Theorem 2.1 to a Higher-Order Equation

Consider the differential equation

$$y^{(n)} = f(x, y, y', \ldots y^{(n-1)}).$$

If the function $f(x, y, y' \ldots y^{(n-1)})$ is analytic at a point $(x_0, y_0, y_1, \ldots y_{n-1})$, that is, it has a Taylor series expansion in powers of $(x - x_0)$, $(y - y_0)(y' - y_1)\ldots(y^{(n-1)} - y_{n-1})$ valid for

$$|x - x_0| < r, \quad |y - y_0| < r, \quad |y' - y_1| < r \ldots \left|y^{(n-1)} - y_{n-1}\right| < r$$

on a $(2r)^{n-1}$ rectangle with center at $(x_0, y_0, y_1 \ldots y_{n-1})$ and if $\left|f(x, y, y' \ldots y^{(n-1)})\right| \leq M$ where M is a positive number, then there is a unique particular solution $y(x)$ that is analytic at $x = x_0$, satisfying the condition

$$y(x_0) = y_0 \quad y'(x_0) = y_1 \dots y^{(n-1)}(x_0) = y_{n-1}$$

which is valid in an interval

$$|x - x_0| < \min\left\{r, \frac{r}{(n+2)M}\right\}.$$

■

Example 2.28

Find the interval of convergence of the Taylor series solution of

$$y'' = y^2 \log x + y'$$

with

$$y(1) = 1, \quad y'(1) = 0, \quad y''(1) = 1.$$

Solution: Choose the rectangle

$$|x - 1| < r, \quad |y - 1| < r, \quad |y' - 0| < r, \quad |y'' - 1| < r$$

with center at (0,1,0,1). Then,

$$|f(x, y)| = |y^2 \log x + y'| < |y^2 \log x| + |y'| < (1 + r)^2 \log(1 + r) + r$$

(from Theorem 2.3).

Thus,

$$M = (1 + r)^2 \log(1 + r) + r.$$

Hence, by Theorem 2.4, the Taylor series solution is convergent in the interval

$$I : |x - 1| < \min\left\{r, \frac{r}{5M}\right\}.$$

Note: To this point, we found the series solution to ODEs at a finite point. Sometimes, we may be interested in finding solutions at a point at infinity. These solutions can be obtained using the method described in Chapter 3.

Exercise Problems

1. Apply the power series method to solve

$$y'' + x^2 y = 2 + x + x^2$$

about $x = 0$.

2. Solve

$$y'' - xy' + 2y = e^{-x}, \quad y(0) = 2, \quad y'(0) = 3$$

3. Consider the Mathieu equation

$$y'' + (3 + 4\cos x)\, y = 0$$

together with the initial conditions

$$y(0) = 1, \quad y'(0) = 3$$

Find the first three nonvanishing coefficients in the power series solution of the Mathieu equation.

4. Find a polynomial solution to

$$(x^2 + 1)y'' - 2xy' + 2y = 0$$

using the power series method.

5. Hermite's equation is

$$y'' - xy' + 2y = e^{-x}$$

Find a power series solution about $x = 0$. Also find polynomial solutions when (i) $p = 0$, (ii) $p = 1$, (iii) $p = 2$. Over what interval is the power series solution guaranteed to be valid?

6. Solve

$$(1 - x^2)y'' - 2xy' + 2y = 0$$

near $x = 0$.

7. Solve

$$y'' - \frac{2}{(1-x)^2}\, y = 0$$

near $x = 0$.

8. Using the Frobenius method, solve

$$x^2 y'' + xy' + \left(x^2 - 4\right) y = 0$$

9. Find two linearly independent solutions of

$$x^2 y'' - xy' + (1 - x) y = 0$$

10. Find the general solution of

$$xy'' + e^{-x} y = 0$$

11. Find the general solution of $xy'' + \lambda y' - y = 0$ when λ is a real parameter in the following cases:
    (i) λ is not an integer
    (ii) $\lambda = 1$
    (iii) $\lambda = 1/2$
12. Use the Frobenius method to find a solution near $x = 4$ for

$$y'' - \frac{1}{x-4} y = 0$$

13. Find, by the Taylor series method, a particular solution of

$$y' = x^2 + y^2, \quad y(0) = 1$$

Also find the interval of validity of the solution.
14. Find a series solution, using the Taylor method, of

$$y' = \cos x + \sin y, \quad y(0) = 0$$

(up to three nonzero terms).
15. Solve

$$y'' = \cos x + \sin y, \quad y(0) = 0, \quad y'(0) = 1$$

Hints:

2. Expand e^{-x} and compare the coefficient of x^n on both sides to obtain the recurrence relation.

$$a_0 = 2, \quad a_1 = 3$$

3. Expand cos x about $x = 0$.

7. The given equation

$$(1-2x+x^2)y''-2y=0, \quad y_1=\frac{1}{1-x}, \quad y_2=(1-x)^2$$

9.

$$y_1(x)=a_0\sum_{n=0}^{\infty}\frac{1}{(n!)^2}x^{n+1};\ y_2(x)=y_1\log x+2x^2-\frac{3}{4}x^3-\frac{11}{108}x^4\ldots$$

10. Write the given equation as

$$x^2y''+xe^{-x}y=0$$

12.

$$y=a_0\left(x-4\right)$$

13.

$$y(x)=1+x+x^2+\frac{4}{3}x^3+\ldots, I=\min\left\{r,\frac{r}{3(2r^2+2r+1)}\right\}, r=\frac{1}{\sqrt{2}}$$

14.

$$y(x)=x+\frac{x^2}{2}-\frac{x^4}{24}+\ldots, \quad \left|\cos x+\sin y\right|<2, \quad I:\left|x\right|<r/6$$

where r is arbitrary.

15.

$$y(x)=x+\frac{x^2}{2}+\frac{x^3}{6}+\ldots, \quad I:\left|x\right|<r/8$$

where r is arbitrary.

Applications

A.L. Shuvalov. The Frobenius power series solution for cylindrically anisotropic radially in homogeneous elastic materials. *Quarterly Journal of Mechanics and Applied Mathematics*, Vol. 56, No. 3, pp. 327–346, 2003.

K.W. Tomantschger. Series solution of coupled differential equations with one regular singular point. *Journal of Computational and Applied Mathematics*, Vol. 140, No. 1–2, pp. 773–783, 2002.

T. Kitamoto. Solution of a linear differential equation in the form of power series and its application. In *Computer Mathematics. Lecture Note Series on Computing*, Vol. 9, edited by K. Shirayanagi and K. Yokoyama. River Edge, NJ: World Scientific, 2001, pp. 45–55.

H.I. Qaisi. A power series approach for the study of periodic motion. *Journal of Sound and Vibrations*, Vol. 196, No. 4, pp. 401–406, 1996.

D.L. Littlefield and P.V. Desai. Frobenius analysis of higher order equations: Incipient buoyant thermal convection. *SIAM Journal on Applied Mathematics*, Vol. 50, No. 6, pp. 1752–1763, 1990.

J. Van Ekeren. *A Treatise on the Hydrogen Bomb*. Hamilton, NZ: University of Waikoto, 2008.

P. Koscik and A. Kopinska. Application of the Frobenius method to the Schrodinger equation for a spherically symmetric potential: An harmonic oscillator. *Journal of Physics A: Mathematics and General*, Vol. 38, pp. 7743–7755, 2005.

R. Ballarini and P. Villaggio. Frobenius method for curved cracks. *International Journal of Fracture*, Vol. 139, pp. 59–69, 2006.

Bibliography

T.M. Apostol. *Mathematical Analysis*. 2nd edition. Boston: Addison-Wesley, 1974.

W.W. Bell. *Special Functions for Scientists and Engineers*. Mineola, NY: Dover, 2004.

M.D. Greenberg. *Advanced Engineering Mathematics*. 2nd edition. Boston: Pearson Education, 1998.

E. Kreyzig. *Advanced Engineering Mathematics*. 9th edition. New Delhi: Wiley India, 2005.

A.H. Nayfeh. *Perturbation Methods*. New York: Wiley, 1972.

R.B. Paris and A.D. Wood. *Asymptotics of High Order Differential Equations*. New York: Wiley, 1986.

M.D. Raisinghania. *Ordinary and Partial Differential Equations*. New Delhi: S. Chand, 2008.

W. Rudin. *Principles of Mathematical Analysis*. 2nd edition. New York: McGraw-Hill, 1976.

G.F. Simmons. *Differential Equations with Applications and Historical Notes*. 2nd edition. Noida, India: Tata McGraw-Hill, 2003.

M. Tenenbaum and H. Pollard. *Ordinary Differential Equation*. Mineola, NY: Dover, 1985.

D.G. Zill. *A First Course in Differential Equations with Modeling Applications*. 9th edition. Stamford, CT: Brooks/Cole, Cengage Learning, 2008.

3
Asymptotic Method

Introduction

Chapter 2 discussed the power series method to find an approximate analytical solution to ordinary differential equations (ODEs). We observed that the series solutions obtained were all convergent in nature, although in some cases, the region of convergence is small. But, the major drawback of it is that the method fails to provide solutions to equations at irregular singular points (IRSPs) and solutions at infinity. Note that, as such, there is no general theory for solution of an ODE about an IRSP (Paris and Wood, 1986). However, the method presented in this chapter—the asymptotic method—provides solutions to problems with irregular singularity at infinity. This method can also be used to provide solutions at infinity to differential equations and to the so-called singularly perturbed problems, which are discussed in detail in Chapter 4. This chapter is limited to the applications of this method to linear differential equations. Refer to the work of Bayat et al. (2012) for applications of the method to nonlinear differential equations. This chapter also provides a method for solving equations involving large parameters.

Another important point to note here regarding the series solution provided by the asymptotic method is that the series solution need not be a convergent series always. But, the characteristic feature of these solutions is that, in spite of their divergent nature, there will be a particular partial sum that provides the best approximation to the solution function of the differential equation considered. It has been shown that these solutions provide a way for computing the numerical values of solution functions of certain differential equations that is otherwise not possible, for example, the values of Bessel's functions at infinity (Nayfeh, 1972).

The chapter is organized as follows: The method of finding asymptotic series solutions at IRSPs at infinity and how to construct solutions valid for large values of the independent variable are covered in Section 3.2. In Section 3.3, we present the asymptotic solutions of regular as well as singularly perturbed problems, followed by the asymptotic method for finding solutions for equations containing a large parameter. A list of applications and bibliography are provided.

Asymptotic Solutions at Irregular Singular Points at Infinity

Before the discussion of the asymptotic method, let us discuss the order notation of a function:

o notation (Small o notation): We say that

$$f(x) = o(g(x))$$

that is, $f(x)$ is small-order $g(x)$ as $x \to 0$ if

$$\underset{x\to 0}{Lt} \frac{f(x)}{g(x)} = 0.$$

O notation (big O notation): We say that

$$f(x) = O(g(x))$$

that is, $f(x)$ is big-order $g(x)$ as $x \to 0$ if there is a positive number M such that $|f(x)| \leq M|g(x)|\, \forall x$ in the neighborhood of zero.

For example,

$$x = o(x^{1/3})$$

as

$$\underset{x\to 0}{Lt} \frac{x}{x^{1/3}} = \underset{x\to 0}{Lt}\, x^{2/3} = 0$$

$$\sin x = O(x) \quad \text{as} \quad |\sin x| \leq M|x|$$

for $x \to 0$ where $M = 1$.

Method of Finding Solutions at Irregular Points

Consider a second-order linear differential equation given by

$$y'' + P(x)\,y' + Q(x)\,y = 0. \tag{3.1}$$

To classify the singularity at infinity, we follow the definitions given in the work of Nayfeh (1972).

For this, first express $P(x)$ and $Q(x)$ in descending powers of x as

$$P(x) = p_0 x^{\alpha} + \ldots \quad \text{and} \quad Q(x) = q_0 x^{\beta} + \ldots \tag{3.2}$$

Now, we classify the point at infinity as follows:

If $\beta \le -4$ and either $\alpha = -1$ with $p_0 = 2$ or $\alpha \le -2$, then the point at infinity is called an ordinary point.

If these conditions are not satisfied, and $\alpha \le -1$ and $\beta \le -2$, then the point at infinity is an RSP.

If one or both of the inequalities $\alpha > -1$, $\beta > -2$ is satisfied, the point at infinity is an IRSP.

Now, let us see how to find the asymptotic solution for a differential equation as in Equation (3.1) at an IRSP of it.

Consider the situation where $P(x)$ and $Q(x)$ are both independent of x. In this case, we assume the solution to Equation (3.1) is

$$y(x) = e^{\lambda x} x^{\sigma} \sum_{n=0}^{\infty} c_n x^{-n}. \tag{3.3}$$

Here, λ is a root of

$$\lambda^2 + p_0\lambda + q_0 = 0 \tag{3.4}$$

with p_0 and q_0 as defined in Equation (3.2) and

$$\sigma = -\frac{\lambda p_1 + q_1}{2\lambda + p_0}. \tag{3.5}$$

The coefficient c_n's are obtained using the recurrence relation that is found by substituting $y(x)$ from Equation (3.3) into the given differential equation.

Let us use examples to illustrate this method.

Example 3.1

Consider the equation

$$y'' + y = 0$$

Solution: Here

$$P(x) = 0; \quad Q(x) = 1$$

Arranging $P(x)$ and $Q(x)$ in descending powers of x and collecting the coefficients, we obtain

$$p_0 = 0; \quad \alpha = 0$$

and

$$q_0 = 1; \quad \beta = 0; \quad q_1 = 0\ldots$$

Using Equation (3.4), we obtain $\lambda = \pm i$, and using Equation (3.5), we obtain $\sigma = 0$.

Thus, the solution for $\lambda = i$ is

$$y(x) = e^{ix} \sum_{n=0}^{\infty} c_n x^{-n}.$$

Computing $y'(x)$, $y''(x)$ and substituting in the given equation, the recurrence relation to find c_n's is

$$c_n = \frac{n+1}{2i} c_{n-1} \quad \text{for} \quad n = 1, 2, 3 \ldots$$

Thus,

$$c_1 = -ic_0; \quad c_2 = \frac{3}{2i} c_1 = -\frac{3}{2} c_0; \quad c_3 = \frac{3}{2i} c_1 = 3ic_0$$

and so on.

The solution is

$$y(x) = e^{ix} c_0 \left(1 - ix^{-1} - \frac{3}{2} x^{-2} + 3ix^{-3} + \ldots \right).$$

Similarly, for $\lambda = -i$, the solution is

$$\bar{y}(x) = e^{-ix} c_0 \left(1 + ix^{-1} - \frac{3}{2} x^{-2} - 3ix^{-3} + \ldots \right).$$

Now, we can construct two real solutions to the given problem by defining

$$u(x)=\frac{y+\bar{y}}{2} \quad \text{and} \quad v(x)=\frac{y-\bar{y}}{2i}.$$

(It is left to the reader to find the two real solutions to the problem.)

Let us now continue with a relatively complicated situation where $P(x)$ and $Q(x)$ are functions of x: Find $p_0, p_1, \ldots, \alpha$ and $q_0, q_1, \ldots, \beta$ by expressing the functions $P(x)$ and $Q(x)$ in ascending powers of x as shown in Equation (3.2).

Now, assume the solution is

$$y(x)=e^{\Lambda(x)}x^{\sigma}u(x) \tag{3.6}$$

where $\Lambda(x)$ is taken to be a polynomial in $x^{m/n}$ and $u(x)=O(1)$ as $x \to \infty$.

We find a constant given by

$$\nu=\alpha+1 \quad \text{or} \quad 2\nu=\beta+2 \tag{3.7}$$

whichever gives the greater value.

If ν is an integer, then the solution given in Equation (3.6) is called a normal solution. Otherwise, it is called a subnormal solution in which $\Lambda(x)$ is a polynomial in $x^{1/2}$ and $u(x)$ is an ascending series in $x^{-1/2}$.

Again, $y(x)$ is constructed and the constants involved are found by comparing the coefficients of different powers of x obtained by substituting $y(x)$ in the given differential equation. If $\nu = 1$, then the constants involved can be found using the formulae given in Equations (3.4) and (3.5).

Example 3.2

Consider the equation

$$y''-xy=0$$

Solution: Here

$$P(x)=0; \quad Q(x)=-x$$

Arranging $P(x)$ and $Q(x)$ in descending powers of x and collecting the coefficients, we obtain

$$p_0=0; \quad \alpha=0$$

and

$$q_0 = -1; \quad \beta = 1; \quad q_1 = 0\ldots.$$

Because the $\alpha > -1$ and $\beta > -2$ conditions are satisfied, the point at infinity is an IRSP for the given differential equation.

Here, $\nu = 3/2$ in view of Equation (3.7). Thus,

$$\Lambda(x) = e^{\lambda_3 x^{3/2} + \lambda_2 x^{1/2}}$$

and

$$u(x) = 1 + a_1 x^{-1/2} + a_2 x^{-2/2} + \ldots.$$

Substituting $y(x)$ in the given equation and comparing the coefficients of like powers of x, we obtain $y(x) = e^{\pm\frac{2}{3}x^{3/2}} x^{-1/4}(1 + O(x^{-3/2}))$. (The reader may verify this.)

A point to note here is that this Airy's equation can be solved much more easily using the WKB method discussed in Chapter 6 of this book.

Asymptotic Method for Constructing Solutions that Are Valid for Large Values of the Independent Variable

The method to construct solutions that are valid for large values of the independent variable (i.e., the solutions valid at infinity) is the same as that of finding solutions at IRSP at infinity. Note that the point at infinity need not be an IRSP for the method to be valid (Nayfeh, 1993). An example illustrates it.

Example 3.3

Consider the equation

$$y'' + \frac{1}{x} y' + y = 0$$

(In literature, this equation is called Bessel's equation of order zero.)

Let us find the asymptotic solution to this equation.

Solution: Here

$$p_0 = 0; p_1 = 1; \alpha = -1 \quad \text{and} \quad q_0 = 1; \beta = 0; q_1 = 0.$$

Thus, the point at infinity is an RSP.

Further, $\nu = \alpha + 1$ or $2\nu = \beta + 2$ takes the value $\nu = 1$. Hence, using Equation (3.4) we obtain $\lambda = \pm i$, and using Equation (3.5), we obtain $\sigma = -\frac{1}{2}$. Thus, for $\lambda = i$,

$$y(x) = e^{ix} x^{-1/2} \sum_{n=0}^{\infty} c_n x^{-n}.$$

Computing $y'(x)$, $y''(x)$ and substituting in the given equation, the recurrence relation to find c_n's is

$$c_{n+1} = -i \frac{\left(n + \frac{1}{2}\right)^2}{2(n+1)} c_n \quad \text{for} \quad n = 1, 2, 3 \ldots$$

The solution is

$$y(x) = e^{ix} x^{-1/2} c_0 \left(1 - i \frac{1}{4.2} x^{-1} - \frac{1.3^2}{4^2.2^2 2!} x^{-2} + i \frac{1.3^2.5^2}{4^3 2^3 3!} x^{-3} \ldots\right)$$

Similarly, for $\lambda = -i$, the solution is

$$\bar{y}(x) = e^{-ix} x^{-1/2} c_0 \left(1 + \frac{1}{4.2} i x^{-1} - \frac{1.3^2}{4^2.2^2 2!} x^{-2} - i \frac{1.3^2.5^2}{4^3 2^3 3!} x^{-3} \ldots\right)$$

Now, we can construct two real solutions to the given problem by defining

$$u(x) = \frac{y + \bar{y}}{2} = x^{-1/2} \left(w_1 \cos x + w_2 \sin x\right)$$

and

$$v(x) = \frac{y - \bar{y}}{2i} = x^{-1/2} \left(w_1 \sin x - w_2 \cos x\right)$$

where

$$w_1(x) = 1 - \frac{1.3^2}{2^2 4^2 2!} x^{-2} + \ldots$$

and

$$w_2(x) = \frac{1}{4.2} x^{-1} - \frac{1.3^2.5^2}{4^3 2^3 3!} x^{-3} \ldots.$$

Thus, the asymptotic solution of Bessel's equation is

$$y(x) = Au(x) + Bv(x).$$

Asymptotic Solutions of Perturbed Problems

Mathematical models of many problems in physics often lead to differential equations involving small parameters. Usually, these parameters have some physical significance and, although very small, cannot be neglected as they have a significant effect on the problem. For example, in the problems related to boundary-layer theory, this parameter could be the nondimensional ratio of the boundary-layer thickness to the typical length scale of the problem. In literature, these problems are termed perturbation problems (refer to Chapter 4 for more details). Many times, the terms involving these parameters make the differential equations so complicated that there is great difficulty in finding the exact analytical solutions to those problems.

Hence, to solve these problems, sometimes simpler models were developed either by letting the parameter zero or by restricting the domain of the problem. But, the solutions thus obtained may or may not tend uniformly to the solution of the given problem. In view of this, the perturbation technique was introduced, by which we start with a simple model obtained by letting the small parameter zero and add corrections or perturbations to attain the solution of the given problem. This perturbation method is detailed in Chapter 4. But, these methods help us to construct solutions that are valid for only small values of the independent variable.

The asymptotic method described in this chapter is much similar to this technique but the crucial point in the method is that the correction terms are found in such a way that the order of these correction terms do not exceed the *order* of the solution of the original or simpler model. Also, the solution thus obtained is valid for large values of the independent variable.

Let us now see how to construct asymptotic solutions to the equations of the form

$$L(y)+\varepsilon N(y)=0 \tag{3.8}$$

where $L(y)=0$ is a linear differential equation that can be readily solved for its solution $y_0(x)$. [In literature, equations of the form as in Equation (3.8) are termed regular perturbation problems.]

We are now to find a solution $y(x,\varepsilon)$ to Equation (3.8) so that, as $\varepsilon \to 0$, $y(x,\varepsilon) \to y_0(x)$. Having determined y_0, we assume that the solution to Equation (3.8) is

$$y(x,\varepsilon) = y_0(x) + \varepsilon y_1(x) + \varepsilon^2 y_2(x) + \ldots. \tag{3.9}$$

Substituting $y(\varepsilon)$ from Equation (3.9) into Equation (3.8) and collecting the coefficients of like powers of ε, we obtain a set of differential equations that determine y_1, y_2, and so on.

Now, the solution of Equation (3.8) is given by Equation (3.9). The arbitrary constants in the solution can be obtained from the given initial or boundary conditions.

Let us examine the example that follows.

Example 3.4

Consider the equation

$$y'' + 2\varepsilon y' + y = 0, y(0) = 0, y'(0) = 1 \tag{3.10}$$

(This equation models the problem of a linear oscillator.)

Let us find a solution using the asymptotic method described previously.

Solution: The problem can be rewritten as

$$(y'' + y) + \varepsilon(2y') = 0$$

Here,

$$L(y) = y'' + y$$

and

$$N(y) = 2y'$$

Assume the solution to Equation (3.10) is $y(x) = y_0(x) + \varepsilon y_1(x) + \varepsilon^2 y_2(x) + \ldots$.

Substituting $y(x)$ in the given equation, we have

$$\left(y_0'' + \varepsilon y_1'' + \varepsilon^2 y_2'' + \ldots\right) + 2\varepsilon\left(y_0' + \varepsilon y_1' + \varepsilon^2 y_2' + \ldots\right) + \left(y_0 + \varepsilon y_1 + \varepsilon^2 y_2 + \ldots\right) = 0$$

$$\left(y_0 + \varepsilon y_1 + \varepsilon^2 y_2 + \ldots\right)(0) = 0,$$

$$\left(y_0' + \varepsilon y_1' + \varepsilon^2 y_2' + \ldots\right)(0) = 1.$$

Comparing the coefficients of different powers of ε, we obtain the following:

Coefficient of ε^0 is

$$y_0'' + y_0 = 0, \quad y_0(0) = 0, \quad y_0'(0) = 1$$

which is the simplified form of the original problem [obtained by putting $\varepsilon = 0$ in Equation (3.10)].

Solving this equation together with the initial conditions, we obtain

$$y_0(x) = \sin x.$$

Now, the coefficient of ε^1 is

$$y_1'' + y_1 = -2y_0', \quad y_1(0) = 0, \quad y_1'(0) = 0.$$

Using the expression for $y_0(x)$ and solving for $y_1(x)$, we obtain $y_1(x) = -x\sin(x)$.

Thus, the solution to the problem is

$$y(x) = \sin x - \varepsilon\, x \sin x + \dots . \tag{3.11}$$

It can be seen that for $|x| \le x_0$ (where x_0 is some positive constant), the correction term $y_1(x)$ is bounded and small for small ε. But, for large x, although ε is small, this correction factor becomes large. This term is called a secular term, whose presence will not allow the solution given by Equation (3.11) approach the solution of the simplified equation for large x.

To obtain a solution valid in the entire range, we use the Lindstedt–Poincaré technique. Refer to the work of Nayfeh (1993) for more details.

Introduce the transformation given by $x = \frac{\xi}{\omega}$ where

$$\omega = 1 + \varepsilon\omega_1 + \varepsilon^2\omega_2 + \dots .$$

Then, $d\xi = \omega dx$ and hence

$$y' = \omega \frac{dy}{d\xi}$$

and

$$y'' = \omega^2 \frac{d^2 y}{d\xi^2}.$$

Let the solution of Equation (3.10) be

$$y(x) = y_0(x) + \varepsilon y_1(x) + \varepsilon^2 y_2(x) + \dots .$$

Now, Equation (3.10) in terms of the new variable ξ is

$$\omega^2 \frac{d^2 y}{d\xi^2} + 2\varepsilon\omega \frac{dy}{d\xi} + y = 0$$

with

$$y(0) = 0, \quad y'(0) = 1.$$

Substituting the expansions for ω and $y(\xi)$ in this equation, we have

$$\left(1 + \varepsilon\omega_1 + \varepsilon^2\omega_2 + \dots\right)^2 \left(y_0''(\xi) + \varepsilon y_1''(\xi) + \varepsilon^2 y_2''(\xi) + \dots\right)$$
$$+ 2\varepsilon\left(1 + \varepsilon\omega_1 + \varepsilon^2\omega_2 + \dots\right)\left(y_0'(\xi) + \varepsilon y_1'(\xi) + \varepsilon^2 y_2'(\xi) + \dots\right)$$
$$+ \left(y_0(\xi) + \varepsilon y_1(\xi) + \varepsilon^2 y_2(\xi) + \dots\right) = 0$$

Collecting the coefficients of

$$\varepsilon^0 : y_0'' + y_0 = 0, \quad y_0(0) = 0, \quad y_0'(0) = 1$$

Solving, $y_0(\xi) = \sin\xi$

$$\varepsilon^1 : y_1'' + y_1 = -2\cos\xi + 2\omega_1 \sin\xi, \quad y_1(0) = 0, \quad y_1'(0) = 0$$

and its solution is

$$y_1(\xi) = \cos\xi - \omega_1\xi\cos\xi - \cos^3\xi + \omega_1 \sin\xi - \sin\xi \sin 2\xi.$$

We see that this solution contains a secular term $\xi \cos \xi$. To remove it, take $\omega_1 = 0$. Now, the correction term is

$$y_1(\xi) = \cos\xi - \cos^3\xi - \sin\xi \sin 2\xi$$

which simplifies to

$$-\frac{\sin\xi \sin 2\xi}{2}.$$

Thus, the solution to the given equation

$$y(\xi) = \sin\xi - \varepsilon\frac{\sin\xi \sin 2\xi}{2} + \dots$$

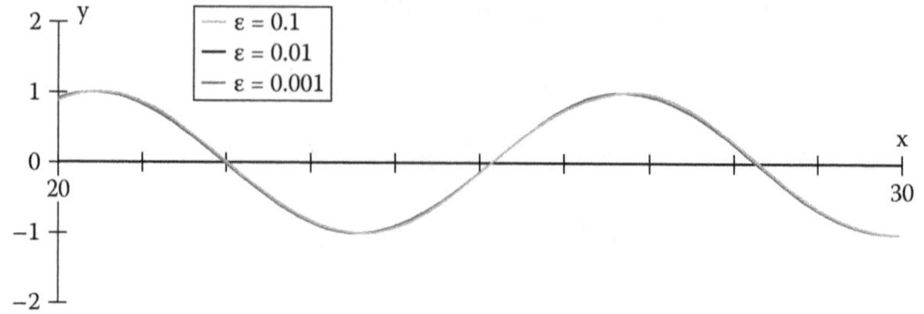

Figure 3.1 Plot of $y_0(x)$ and $y(x)$ for large x and $\varepsilon = 0.001$, 0.01, and 0.1.

that is,

$$y(x) = \sin x - \varepsilon \frac{\sin x \sin 2x}{2} + \ldots.$$

We can observe from the graph in Figure 3.1 that, for small ε, the first partial sum itself approximates the solution $y_0(x)$ of the simple equation.

The technique described is a simple substitution method that could remove the secular terms in the solution of the differential equation in which the small parameter is present as a coefficient of a term other than the highest derivative term. But, if the small parameter is a coefficient of the highest derivative term of a differential equation, the problems are called singular perturbation problems. For these problems, it may not always be possible to find a simple substitution, as given in the example, to suppress the secular terms in the solution to make them valid for large values of the independent variable. Hence, to address these situations, several methods have been proposed, such as the following:

1. Boundary-layer method
2. Multiple-scale method
3. WKB method

and so on. These methods are described in subsequent chapters.

Solutions to ODEs Containing a Large Parameter

Let us now see how to find solutions to ODEs containing a large parameter. The method is credited to Horn (1899).

Consider the differential equation of the form

$$y'' + q(x,\lambda)\,y = 0 \tag{3.12}$$

where

$$q(x,\lambda) = \lambda^{2k}\sum_{n=0}^{\infty}\lambda^{-n}q_n(x) \quad \text{as } \lambda \to \infty \tag{3.13}$$

with $q_0 \neq 0$ in the interval of interest, and k is a positive integer.

The asymptotic solution of this problem is taken in either of the following forms:

$$y(x) = \left(\sum_{n=0}^{\infty}\lambda^{-n}a_n(x)\right)e^{\lambda^k\sum_{n=0}^{k-1}\lambda^{-n}g_n(x)} \tag{3.14}$$

$$y(x) = e^{\lambda^k\sum_{n=0}^{\infty}\lambda^{-n}g_n(x)} \tag{3.15}$$

Substituting either of these forms in Equation (3.12) and equating the like powers of λ, we obtain equations that determine the forms of $a_n(x)$ and $g_n(x)$.

Example 3.5

Determine a first approximation to the eigenvalue problem

$$y'' + \lambda^2 f(x)\,y = 0, \quad f(x) > 0$$

for large λ.

Solution: Comparing with Equation (3.13), we have

$$k = 1; \quad q_0(x) = f(x)$$

Thus, taking the form of $y(x)$ from Equation (3.14), we have

$$y(x) = \sum_{n=0}^{\infty}\lambda^{-n}a_n(x)e^{\lambda g_0(x)}$$
$$\approx a_0(x)e^{\lambda g_0(x)}$$

up to the first approximation.

Computing $y''(x)$ and substituting in the given problem, we have

$$a_0''(x)e^{\lambda g_0(x)} + 2\lambda a_0'(x)g_0'(x)e^{\lambda g_0(x)} + \lambda^2 a_0(x)e^{\lambda g_0(x)}\left(g_0'(x)\right)^2$$
$$+\lambda a_0(x)e^{\lambda g_0(x)}g_0''(x) + \lambda^2 a_0(x)f(x)e^{\lambda g_0(x)} = 0$$

Comparing the coefficients of different powers of λ, we obtain

$$g_0'^2 + f(x) = 0$$
$$2g_0'a_0' + g_0''a_0 = 0$$

Solving these equations gives

$$g_0 = \pm i\int\sqrt{f(x)}dx$$

and

$$a_0 = c\int\left(f(x)\right)^{-1/4} dx$$

and thus the asymptotic solution up to the first approximation is

$$y(x) \approx c\left(f(x)\right)^{-1/4} e^{\pm i\lambda\int\sqrt{f(x)}dx}.$$

Exercise Problems

1. Use the method described in the previous section to find a solution to the equation

$$y'' + y + \varepsilon y^3 = 0, \quad y(0) = 1, \quad y'(0) = 0$$

   which is valid for large x.
2. Find the solution to the eigenvalue problem

$$y'' + \lambda^2 y = 0$$

   which is valid for large values of λ.
3. Determine a first approximation to the eigenvalue problem

$$y'' + \lambda^2 f(x)y = 0, \quad f(x) > 0$$

   for large λ [using the form in Equation (3.15)].

Hints:

1.

$$y_0(\xi) = \cos\xi$$

$$y_1(\xi) = \frac{\cos 3\xi - \cos\xi}{32}; \quad \omega_1 = 3/8$$

2. See Example 3.5 ($f(x) = 1$).
3.

$$g_0 = \pm i \int \sqrt{f(x)}dx; \quad g_1 = -\frac{1}{2}\ln g_0'$$

Applications

The following references give greater insight to the method explained in the chapter:

M. Jiaqi and L. Wantao. Asymptotic solution of singularly perturbed problem for a nonlinear singular equation. *Applied Mathematics—A Journal of Chinese Universities, Series B*, Vol. 19, No. 2, pp. 187–190, 2004.

G. Nistor. Asymptotic solution of the Cauchy problem for the Prandtl equation. *ROMAI Journal*, Vol. 2, No. 2, pp. 123–128, 2006.

A.G. Petrov. Asymptotic methods for solving the Hamilton equations with the use of a parametrization of canonical transformations. *Differential Equations*, Vol. 40, No. 5, pp. 672–685, 2004.

Bibliography

M. Bayat, I. Pakar, and G. Domairry. Recent developments of some asymptotic methods and their applications for nonlinear vibration equations in engineering problems: A review. *Latin American Journal of Solid and Structures*, Vol. 9, pp. 145–234, 2012.

O.M. Bender and S.A. Orszag. *Advanced Mathematical Methods for Scientists and Engineers*. New York: McGraw-Hill.

A.G. Birkhoff. Quantum mechanics and asymptotic series. *Bulletin of the American Mathematical Society*, Vol. 39, pp. 681–700, 1933.

F.T. Cope. Formal solutions of irregular linear differential equations I. *American Journal of Mathematics*, Vol. 56, pp. 411–437, 1934.

F.T. Cope. Formal solutions of irregular linear differential equations II. *American Journal of Mathematics*, Vol. 58, pp. 130–140, 1936.

E.T. Copson. *Asymptotic Expansions*. Cambridge, UK: Cambridge University Press, 1965.

J. Cousteix and J. Mauss. *Asymptotic Analysis and Boundary Layers*. New York: Springer, 2007.

A. Erdélyi. *Asymptotic Expansions*. New York: Dover, 1965.

J. Horn. Ueber eine lineare Differentialgleichug zweiter Ordnung mit einem willkurlichen Parameter. *Mathematische Annalen*, Vol. 52, pp. 271–292, 1899.

J. Kevorkian and J.D. Cole. *Perturbation Methods in Applied Mathematics*. New York: Springer, 1968.

R.P. Kuzmina. *Asymptotic Methods for Ordinary Differential Equations*. Dordrecht, the Netherlands: Kluwer Academic.

R.E. Langer. The asymptotic solutions of ordinary linear differential equations of the second order, with special reference to a turning point. *Transactions of the American Mathematics Society*, Vol. 67, pp. 461–490, 1949.

M. Shkil. On asymptotic methods in the theory of differential equations of mathematical physics. *Nonlinear Mathematical Physics*, Vol. 3, pp. 40–50, 1996.

A.H. Nayfeh. *Perturbation Methods*. New York: Wiley, 1972.

A.H. Nayfeh. *Introduction to Perturbation Techniques*. New York: Wiley, 1993.

R.W. Ogden and D.Y. Gao. *Introduction to Asymptotic Methods*. Boca Raton, FL: CRC Press, 2006.

F.W.J. Olver. The asymptotic solution of linear differential equations of the second order for large values of a parameter. *Philosophical Transactions*, 247, pp. 307–327, 1954.

R.B. Paris and A.D. Wood. *Asymptotics of High Order Differential Equations*. New York: Wiley, 1986.

S.Y. Slavyanov. *Asymptotic Solutions of the One-Dimensional Schrödinger Equation*. Providence, RI: American Mathematical Society, 1996.

M. Van Dyke. *Perturbation Methods in Fluid Mechanics*. New York: Academic Press, 1964.

B. Vasil'eva. Asymptotic methods in the theory of ordinary differential equations with small parameters multiplying the highest derivatives. *Uspekhi Matematicheskikh Nauk*, pp. 25–231, 1962.

W. Wasow. *Asymptotic Expansions of Ordinary Differential Equations*. New York: Wiley, 1965.

R.B. White. *Asymptotic Analysis of Differential Equations*. Revised edition. London: Imperial College Press, 2010.

4

PERTURBATION TECHNIQUES

Introduction

This chapter describes, in detail, the perturbation technique through which we find solutions to a class of problems called perturbation problems. This method helps to construct solutions to problems involving a small parameter ε, and the solutions so obtained are valid for small values of the independent variable. This method is detailed in Section 4.2. Section 4.3 describes the asymptotic method for finding solutions to another class of perturbation problems, the singular perturbation problems. This method is called the boundary-layer method as mentioned in Chapter 3.

The perturbation theory had its roots in early studies of celestial mechanics, for which the theory of epicycles was used to make small corrections to the prediction of the path of planets. Later, Charles Eugene used it to study the solar system, in particular the earth-sun-moon system. Now, it finds applications in many fields, such as fluid dynamics, quantum mechanics, quantum chemistry, quantum field theory, and so on.

The idea behind the perturbation method is that we start with a simplified form of the original problem (which may be extremely difficult to handle) and then gradually add corrections or perturbations, so that the solution matches that of the original problem. The simplified form is obtained by letting the perturbation parameter take the zero value.

Basic Idea behind the Perturbation Method

As described in Chapter 3, we assume the solution to a perturbation problem is an infinite series in terms of the perturbation parameter. This series is called the perturbation series and is in the form

$$y(x) = y_0(x) + \varepsilon y_1(x) + \varepsilon^2 y_2(x) + \ldots$$

where ε is the perturbation parameter.

Here, $y_0(x)$ is the solution to the simplified form of the original problem, and the other terms describe the deviation in the solution caused by the deviation in the original problem. The terms $y_1(x), y_2(x), y_3(x)$... represent higher-order terms, which are found iteratively using the procedure described in the section that follows.

Let us now discuss the classification of perturbation problems in detail: Based on how a change in perturbation parameter affects the solution of the original equation, problems are classified as

1. Regular perturbation problems; or
2. Singular perturbation problems.

Regular perturbation theory deals with those problems for which a small change of perturbation parameter induces a small change in the solution, whereas singular perturbation theory deals with those problems for which a small change of the perturbation parameter induces a large change in the solution. In the regularly perturbed problems, the small parameter is present in the term/terms other than the one involving the highest-order derivative.

Consider the examples that follow, in which we classify the equation based on the definitions given.

Example 4.1

$$y'' + \varepsilon y' = 1, \quad y(0) = 0, \quad y'(0) = 0$$

The exact solution to this problem is

$$y(x) = \left(\frac{e^{-\varepsilon x} + \varepsilon x - 1}{\varepsilon^2} \right).$$

(It is left to the reader to derive the solution.)

This problem is a regularly perturbed problem, as is illustrated by Figure 4.1. The figure shows the variation of the exact solution of the initial value problem (IVP) in Example 4.1 with the perturbation parameter ε. It can be observed that a small change in ε produces a slight deviation in the solution.

Example 4.2

$$\varepsilon y'' - y' = 1, \quad y(0) = 0, \quad y(1) = 0$$

is a singular perturbation problem.

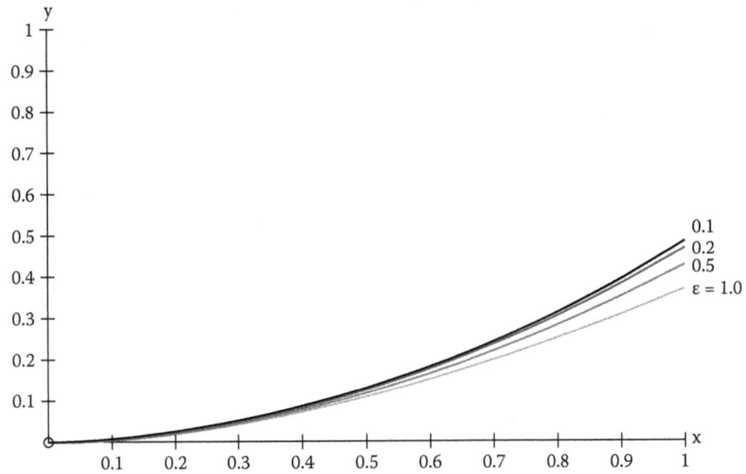

Figure 4.1 Plot of exact solution *y(x)* for different values of ε.

Its exact solution is

$$y(x) = \frac{1 - \varepsilon x - e^{-\varepsilon x}}{\varepsilon^2}.$$

Figure 4.2 shows the variation of the exact solution with the perturbation parameter ε. Here, in contrast to the previous example, it can be observed that a small change in ε produces a great deviation in the solution specifically at the boundary point $x = 1$.

Regular Perturbation Theory

We are all familiar with the Taylor series expansion for an infinitely derivable function $f(x)$ about $x = a$ as

$$f(a+\varepsilon) = f(a) + \varepsilon f'(a) + \frac{\varepsilon^2}{2!} f''(a) + \ldots.$$

In general, for any function $f(x)$, this expansion either fails to converge or fails to capture the behavior of the function. But, for a small change in x, $f(x+\varepsilon)$ always converges to $f(x)$. Using this idea, we construct the solution to the problems that come under the class of regular perturbation problems as discussed next.

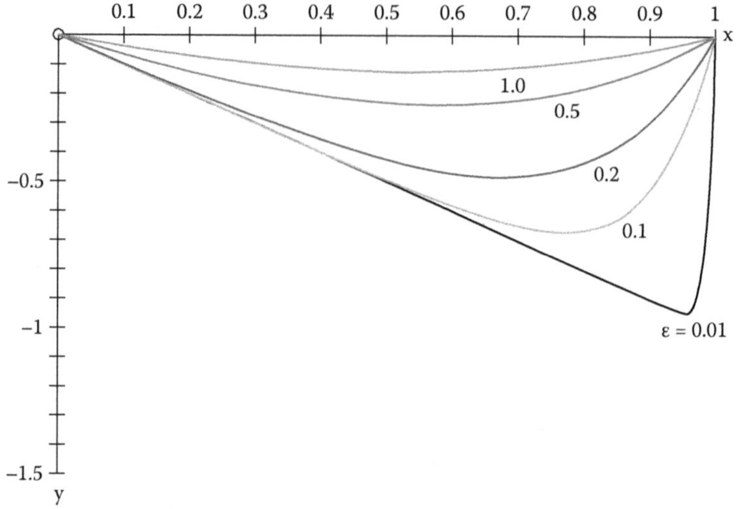

Figure 4.2 Plot of exact solution $y(x)$ for different values of ε.

Solution Method

Step (i): Let $y(x) = y_0(x) + \varepsilon y_1(x) + \varepsilon^2 y_2(x) + \ldots$ be the solution of the given equation, where $y_0(x), y_1(x), y_2(x)\ldots$ are to be determined.

Step (ii): Substitute $y(x)$ in the given equation.

Step (iii): Expand the governing equations as a series of ε, compare the coefficients of the like powers of ε, and solve the resulting differential equations for $y_0(x), y_1(x), y_2(x)\ldots$.

Now, we illustrate the procedure through some examples.

Example 4.3

Solve

$$y'' + \varepsilon y' = 1, \quad y(0) = 1, \quad y'(0) = 0 \tag{4.1}$$

Solution: Assume the solution to Equation (4.1) is

$$y(x) = y_0(x) + \varepsilon y_1(x) + \varepsilon^2 y_2(x) + \ldots$$

Substituting $y(x)$ in the given equation, we have

$$\left(y_0'' + \varepsilon y_1'' + \varepsilon^2 y_2'' + \ldots\right) + \varepsilon\left(y_0' + \varepsilon y_1' + \varepsilon^2 y_2' + \ldots\right) = 1$$

Comparing the coefficients of different powers of ε, we obtain
Coefficient of ε^0:

$$y_0'' = 1, \quad y_0(0) = 1, \quad y_0'(0) = 0$$

which is the simplified form of the original problem [obtained by putting $\varepsilon = 0$ in Equation (4.1)].

Solving the this equation together with the initial conditions, we have

$$y_0(x) = 1 + \frac{x^2}{2}.$$

Now, comparing the coefficient of ε^1, we obtain

$$y_1'' + y_0' = 0, \quad y_1(0) = 0, \quad y_1'(0) = 0.$$

Using the expression for $y_0(x)$ and solving for $y_1(x)$, we obtain

$$y_1(x) = -\frac{x^3}{3!}$$

Similarly, the next coefficient is

$$y_2'' + y_1' = 0, \quad y_2(0) = 0, \quad y_2'(0) = 0$$

Solving this, we obtain

$$y_2(x) = \frac{x^4}{4!}$$

and so on.

Thus, the solution to (4.1) is

$$y(x) = \left(1 + \frac{x^2}{2!}\right) + \varepsilon\left(-\frac{x^3}{3!}\right) + \varepsilon^2\left(\frac{x^4}{4!}\right) + \ldots. \tag{4.2}$$

It can be shown that the term in the first right-hand parentheses of Equation (4.2) is the solution to the simplified form of the original Equation (4.1).

Figure 4.3 presents the exact and the approximate solutions to Equation (4.1) for different values of ε.

Note: This comparison between the exact and the approximate solutions may not always be possible, as shown in the next examples.

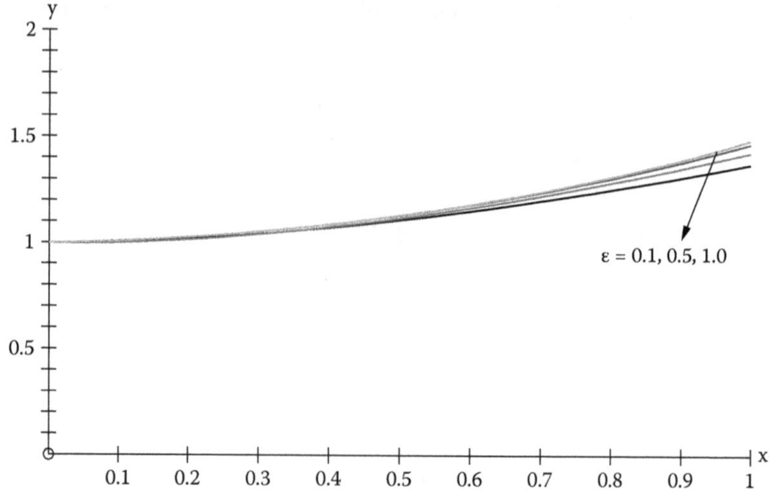

Figure 4.3 Comparison between exact and the approximate solutions.

Example 4.4

Solve

$$y' = \varepsilon e^{y^2}, \quad y(0) = 1 \tag{4.3}$$

Solution: Assume the solution is

$$y(x) = y_0(x) + \varepsilon y_1(x) + \varepsilon^2 y_2(x) + \ldots.$$

Substituting $y(x)$ in the given equation, we have

$$y_0' + \varepsilon y_1' + \varepsilon^2 y_2' + \ldots = \varepsilon\left(e^{\left(y_0+\varepsilon y_1+\varepsilon^2 y_2+\ldots\right)^2}\right)$$

$$= \varepsilon\left(1 + \left(y_0 + \varepsilon y_1 + \varepsilon^2 y_2 + \ldots\right)^2 + \frac{\left(y_0 + \varepsilon y_1 + \varepsilon^2 y_2 + \ldots\right)^4}{2!} + \ldots\right)$$

Comparing the coefficient of ε^0, we obtain $y_0' = 0, y_0(0) = 1$, which is the simplified form of the original problem [obtained by putting $\varepsilon = 0$ in Equation (4.3)]. Solving this, we obtain $y_0 = 1$.

Notice that comparing the coefficient of ε^1, we obtain $y_1' = e^{y_0^2}, y_1(0) = 0$.

Using the expression for y_0 and solving for y_1, we have $y_1 = ex$. Thus, the solution to Equation (4.3) is $y = 1 + \varepsilon ex$ (up to the first approximation).

Example 4.5

Solve

$$y'' + y + \varepsilon y^3 = 0, \quad y(0) = 1, \quad y'(0) = 0 \qquad (4.4)$$

In literature, this equation is referred to as the unforced Duffing equation, named after Georg Wilhelm Christian Casper Duffing. It is used to model hard spring and soft spring oscillations.

Solution: Assume the solution as

$$y(x) = y_0(x) + \varepsilon y_1(x) + \varepsilon^2 y_2(x) + \ldots$$

Substituting $y(x)$ in the given equation, we have

$$(y_0'' + \varepsilon y_1'' + \varepsilon^2 y_2'' + \ldots) + (y_0 + \varepsilon y_1 + \varepsilon^2 y_2 + \ldots)$$
$$+ \varepsilon (y_0 + \varepsilon y_1 + \varepsilon^2 y_2 + \ldots)^3 = 0$$

Comparing the coefficient of ε^0, we have

$$y_0'' + y_0 = 0, \quad y_0(0) = 1, \quad y_0'(0) = 0$$

Thus, $y_0 = \cos x$.

Comparing the coefficient of ε^1, we have

$$y_1'' + y_1 + y_0^3 = 0, \quad y_1(0) = 0, \quad y_1'(0) = 0$$

Using the expression for $y_0(x)$ and solving for $y_1(x)$, we have

$$y_1(x) = \frac{-1}{32}\cos x - \frac{3}{8}x\sin x + \frac{1}{32}\cos 3x$$

Thus, the solution to Equation (4.4) up to the first approximation is

$$y(x) = \cos x + \varepsilon\left(-\frac{1}{32}\cos x + \frac{3}{8}x\sin x + \frac{1}{32}\cos 3x\right) + \ldots$$

Example 4.6

Solve

$$y'' + y + \varepsilon(y^2 - 1)y' = 0, \quad y(0) = 1, \quad y'(0) = 0 \qquad (4.5)$$

Solution: Assume the solution is

$$y(x) = y_0(x) + \varepsilon y_1(x) + \varepsilon^2 y_2(x) + \ldots$$

Substituting *y(x)* in the given equation, we have

$$\left(y''+\varepsilon y_1''+\varepsilon^2 y_2''+\ldots\right)+\left(y_0+\varepsilon y_1+\varepsilon^2 y_2+\ldots\right)$$
$$+\varepsilon\left(\left(y_0+\varepsilon y_1+\varepsilon^2 y_2+\ldots\right)^2-1\right)\left(y'+\varepsilon y_1'+\varepsilon^2 y_2'+\ldots\right)=0$$

Comparing the following:
Coefficient of ε^0, we have

$$y_0''+y_0=0,\ y_0(0)=1, y_0'(0)=0$$

Thus, $y_0=\cos x$
Coefficient of ε^1, we have

$$y_1''+y_1+(y_0^2-1)y_0'=0,\ y_1(0)=0, y_1'(0)=0$$

Using the expression for $y_0(x)$ and solving for $y_1(x)$, we have

$$y_1(x)=-\frac{9}{32}\sin x+\frac{3}{8}x\cos x-\frac{1}{32}\sin 3x$$

Thus, the solution to Equation (4.5) is

$$y(x)=\cos x+\varepsilon\left(-\frac{9}{32}\sin x+\frac{3}{8}x\cos x-\frac{1}{32}\sin 3x\right)+\ldots$$

(up to the first approximation).

Singular Perturbation Theory

The singular perturbation theory deals with the boundary value problems in which the perturbation parameter occurs as the coefficient of the highest-order derivative, and hence a small change in the parameter induces large change in its solution, as is seen in Example 4.2.

Also, when we substitute

$$y(x)=y_0(x)+\varepsilon y_1(x)+\varepsilon^2 y_2(x)+\ldots$$

in the given problem and collect the coefficient of ε^0, we obtain a lower-order differential equation than that of the given one. This is because of the presence of ε as the coefficient of the highest-order derivative. Hence, its solution cannot satisfy all the given initial or boundary conditions. We illustrate this in the next examples.

Example 4.7

Solve

$$\varepsilon y' + y = 1, \quad y(0) = 0$$

Solution: Substituting

$$y(x) = y_0(x) + \varepsilon y_1(x) + \varepsilon^2 y_2(x) + \ldots$$
$$y'(x) = y_0'(x) + \varepsilon y_1'(x) + \varepsilon^2 y_2'(x) + \ldots$$

in the given equation, we obtain

$$\varepsilon(y_0' + \varepsilon y_1' + \varepsilon^2 y_2' + \ldots) + (y_0 + \varepsilon y_1 + \varepsilon^2 y_2 + \ldots) = 1$$

Then, comparison of the coefficient of ε^0 gives $y_0 = 1$ (which does not satisfy the given condition $y_0(0) = 0$).

Hence, we say that there is an initial layer at $x = 0$m which can be seen in Figure 4.4.

Now, to find the required solution, let us introduce a transformation in the form $\xi = \frac{x}{\varepsilon}$. Then, the given equation transforms into

$$\frac{dy}{d\xi} + y = 1, \quad y(0) = 0$$

and its solution is

$$y = 1 - e^{-\xi} = 1 - e^{\frac{-x}{\varepsilon}}$$

which is the required solution to the given IVP for a given ε.

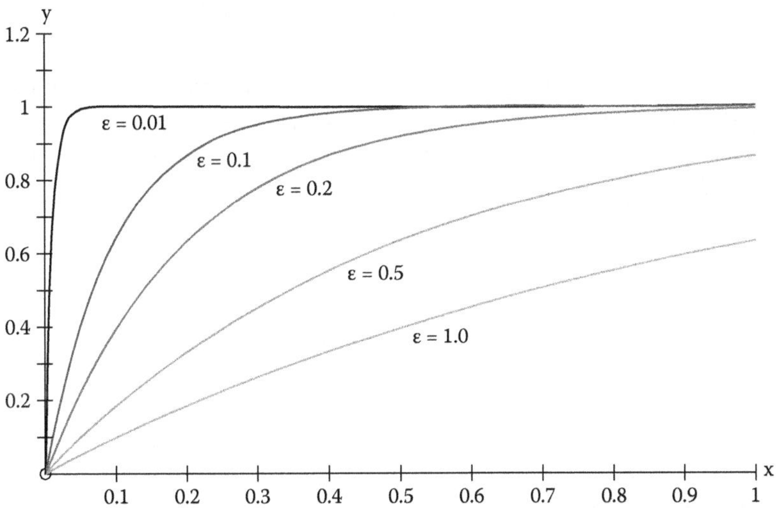

Figure 4.4 Plot of the exact solution for different values of ε.

Example 4.8

Solve

$$\varepsilon y'' - y' = 1, \quad y(0) = 0, \quad y(1) = 0 \tag{4.6}$$

Solution: Put

$$y(x) = y_0(x) + \varepsilon y_1(x) + \varepsilon^2 y_2(x) + \ldots$$

in the given differential equation and the given conditions.

Then, the comparison of the coefficient of ε^0 gives

$$y_0' = -1, \quad y_0(0) = 0, \quad y_0(1) = 0$$

Because this equation is a first-order differential equation and there are two conditions, it is obvious that its solution cannot satisfy both the conditions simultaneously. Hence, we say that there is a boundary layer at the other boundary point where the condition is not satisfied.

For this example, by considering the condition $y_0(0) = 0$, the solution is $y_0(x) = -x$, and the boundary layer is said to be at $x = 1$ (because this solution does not satisfy the condition at $x = 1$). The solution is $y_0(x) = 1 - x$ when $y_0(1) = 0$, and in this case, the boundary layer is said to be at $x = 0$. Hence, in either case, we can expect a boundary layer at the other boundary point.

But, in Figure 4.2, we see that the solution to this problem changed rapidly in the vicinity of the boundary point $x = 1$. Hence, there is a boundary layer at $x = 1$ but not at $x = 0$.

Because it is not always possible to find the boundary layer by plotting the exact solution to the problem, we may have to look for a method that helps identify such a point. One such method is the boundary-layer method (Van Dyke, 1964; Wasow, 1965).

Boundary-Layer Method

Step (i): Finding the outer solution [denoted by $Y_0(x)$]:

Substitute

$$y(x) = y_0(x) + \varepsilon y_1(x) + \varepsilon^2 y_2(x) + \ldots$$

in the given differential equation and collect the coefficient of ε^0. Identify the point (denoted by x_0) where we have a boundary layer.

Let us find the outer solution to the problem given in Equation (4.6). Assume the boundary layer is at $x_0 = 0$. Then, the outer solution is $Y_o(x) = 1 - x$.

Step (ii): Finding the inner solution or the boundary-layer solution [which is denoted by $Y_i(\xi)$ at the point x_0].

For this, we have to use the stretching transformation $\xi = \dfrac{x - x_0}{\varepsilon^{\alpha}}$ and substitute in the given differential equation. Now, find the transformed equation together with the new boundary conditions. Further, use the consistency check to find the value of α, which is as follows: In this consistency check, we find α in such a way that the given singularly perturbed equation becomes a regularly perturbed one.

Let us see the same example given previously to understand this concept.

Because the boundary point is taken to be $x = 0$, consider the transformation $\xi = \dfrac{x}{\varepsilon^{\alpha}}$. Then,

$$\frac{d}{dx} = \frac{d\xi}{dx}\frac{d}{d\xi} = \frac{1}{\varepsilon^{\alpha}}\frac{d}{d\xi}$$

and

$$\frac{d^2}{dx^2} = \frac{1}{\varepsilon^{2\alpha}}\frac{d^2}{d\xi^2}.$$

Substituting these in Equation (4.6), we obtain

$$\varepsilon^{1-2\alpha}\frac{d^2y}{d\xi^2} - \varepsilon^{-\alpha}\frac{dy}{d\xi} = 1, \quad y(\xi = 0) = 0$$

From this equation, we see that, by equating to zero, the different powers of ε, α can take on values ½ or 1. But, only $\alpha = 1$ makes the equation regularly perturbed.

Finally, find the inner solution $Y_i(\xi)$ that is the coefficient of ε^0 in the given equation obtained by substituting

$$y(\xi) = y_0(\xi) + \varepsilon y_1(\xi) + \varepsilon^2 y_2(\xi) + \ldots$$

in it.

For the example, the inner solution is given by

$$\frac{d^2 y_0}{d\xi^2} - \frac{dy_0}{d\xi} = 0, \quad y_0(0) = 0$$

Solving this, we obtain the inner solution as $Y_i(\xi) = y_0 = c(1 - e^{\xi})$, where c is an arbitrary constant.

Step (iii): Matching technique

From the previous two steps, we have found two different solutions, each valid in different parts of the interval described by the two boundary points. Hence, to find a solution valid in the entire interval, we need to impose the matching condition that says that the boundary-layer solution and the solution in the main region (outer solution) must agree or match, for instance, mathematically

$$\operatorname*{Lim}_{x \to x_0} Y_o(x) = \operatorname*{Lim}_{\xi \to \infty} Y_i(\xi).$$

Here, for the example considered, $\operatorname*{Lim}_{x \to 0} Y_o(x) = 1$, whereas $\operatorname*{Lim}_{\xi \to \infty} Y_i(\xi) = c(1 - e^{\xi})$. Hence, we cannot find a c satisfying the machining condition. Thus, we can conclude there is no boundary layer at $x = 0$, which can be also be seen from Figure 4.2.

Step (iv): Composite expansion

If there is a boundary layer at $x = x_0$, we proceed to find the solution that is valid in the entire interval. It is given by

$$y_{comp}(x) = Y_o(x) + Y_i(\xi) - \text{Match}$$

where

$$\text{Match} = \lim_{\xi \to \infty} Y_i(\xi) = \lim_{x \to x_0} Y_O(x).$$

This function $y_{comp}(x)$ is called a composite expansion, and it forms the solution to the given boundary value problem. [The reader may try the solution to the problem given by Equation (4.6) under the assumption that there is a boundary layer at $x = 1$.]

Example 4.9

Solve the boundary value problem $\varepsilon^2 y'' - y = -1$ subject to

$$y(0) = 0, \quad y(1) = 1 \tag{4.7}$$

Solution: An exact solution to this equation is

$$y(x) = \frac{\left(1 - e^{\frac{-x}{\varepsilon}}\right)\left(e^{\frac{x}{\varepsilon}} + e^{\frac{2}{\varepsilon}}\right)}{\left(e^{\frac{2}{\varepsilon}} - 1\right)} \tag{4.8}$$

Note that for more difficult problems, exact solutions are not available. Here, we use this exact solution to check the validity of the method.

We begin with a regular perturbation expansion:

$$y(x) = y_0(x) + \varepsilon y_1(x) + \varepsilon^2 y_2(x) + \ldots$$

and

$$y''(x) = y_0''(x) + \varepsilon y_1''(x) + \varepsilon^2 y_2''(x) + \ldots \tag{4.9}$$

Substituting Equation (4.9) in the given equation and collecting like terms, we obtain

$$\varepsilon^0 : y_0(x) = 1, \quad y_0(0) = 0, \quad y_0(1) = 1$$

$$\varepsilon^1 : y_1(x) = 0, \quad y_1(0) = 0, \quad y_1(1) = 0$$

Hence, the series solution is

$$y(x) = y_0(x) = 1 \tag{4.10}$$

Here, the solution satisfies only one boundary condition given by $y(1) = 1$. Hence, there is a boundary layer near $x = 0$. So, we divide the region of interest $x \in [0, 1]$ into two parts, a thin *inner region* or *boundary layer* around $x_0 = 0$ and an *outer region* elsewhere.

Outer region: The solution in the outer region is $Y_o(x) = 1$ [From Equation (4.10)].

Inner region: In the inner region, we choose a new independent variable ξ defined as $\xi = \frac{x}{\varepsilon^\alpha}$. Substituting ξ in Equation (4.7), we obtain

$$\varepsilon^{1-2\alpha} \frac{d^2 y}{d\xi^2} - y = -1, \; y(\xi = 0) = 0$$

Using a consistency check, we have $\alpha = 1/2$.

After substituting

$$y(\xi) = y_0(\xi) + \varepsilon y_1(\xi) + \varepsilon^2 y_2(\xi) + \ldots$$

in the previous equation, the inner solution is governed by

$$\frac{d^2 y_0}{d\xi^2} - y_0 = -1, \quad Y_0(0) = 0$$

Solving this,

$$y_0(\xi) = c_2(-e^{\xi} + e^{-\xi}) - e^{\xi} + 1 = Y_i(\xi)$$

Using the matching technique, $\lim_{\xi \to \infty} Y_i(\xi) = \lim_{x \to 0} Y_o(x)$, we obtain $c_2 = -1$.

Hence, the inner solution

$$Y_i(\xi) = 1 - e^{-\xi} = 1 - e^{-\frac{x}{\varepsilon}} \tag{4.11}$$

and the

$$\text{Match} = \lim_{\xi \to \infty} Y_i(\xi) = \lim_{x \to x0} Y_o(x)$$

$$\text{Match} = \lim_{\xi \to \infty} (1 - e^{-\xi}) = \lim_{x \to 0} 1 = 1$$

Thus, the composite solution is

$$y_{comp}(x) = Y_i(x) + Y_o(x) - \text{Match}$$

$$y_{comp}(x) = \left(1 - e^{\frac{-x}{\varepsilon}}\right) + 1 - 1 = \left(1 - e^{\frac{-x}{\varepsilon}}\right)$$

and there is a boundary layer at $x = 0$.

Figure 4.5 depicts the boundary layer and the exact solution to the given Equation (4.7) for different values of ε. Figure 4.6 shows the comparison between the exact and the composite solutions of Equation (4.8). It can be seen from Figure 4.5 that the exact and the composite solutions coincide for small values of ε.

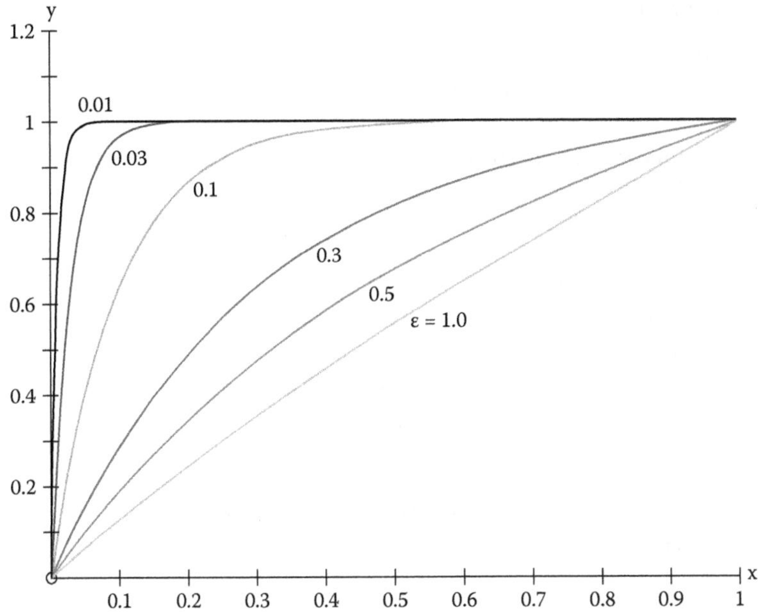

Figure 4.5 Plot of the exact solution of Equation (4.7) for different values of ε.

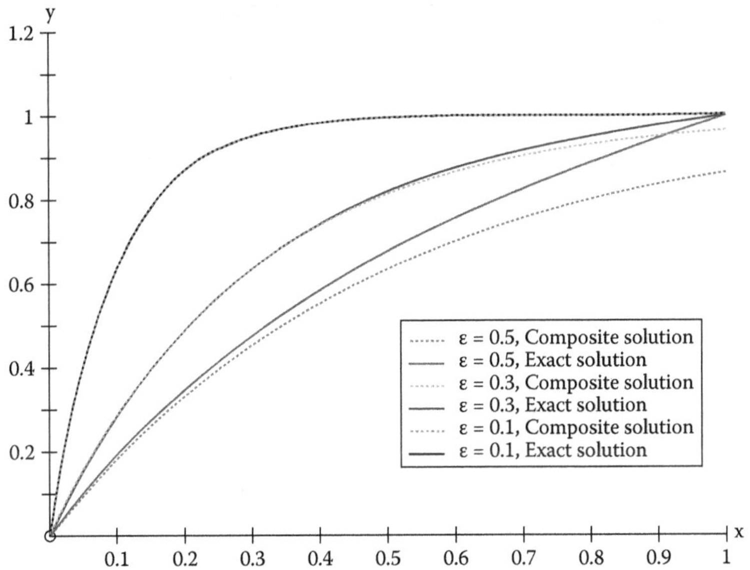

Figure 4.6 Comparison of the exact and the composite solutions.

Example 4.10

Solve

$$\varepsilon y'' - y = 0, \quad y(0) = 1, \quad y(\infty) = 0$$

Solution: Outer solution:
Put

$$y(x) = y_0(x) + \varepsilon y_1(x) + \varepsilon^2 y_2(x) + \ldots$$

in the given equation.

Collecting the coefficient of ε^0, gives $y_0(x) = 0$. Hence, the outer solution is $Y_o(x) = 0$.

This solution satisfies the boundary condition $y(\infty) = 0$. Therefore, there is a boundary layer at $x = 0$.

Inner solution:
Here, $x_0 = 0$. Let $\xi = \frac{x}{\varepsilon^\alpha}$. Then, the given differential equation transforms into

$$\varepsilon^{1-2\alpha} \frac{d^2 y}{d\xi^2} - y = 0. \qquad (4.12)$$

Using a consistency check, we have $\alpha = 1/2$.

Substituting

$$y(\xi) = y_0(\xi) + \varepsilon y_1(\xi) + \varepsilon^2 y_2(\xi) + \ldots$$

in Equation (4.12), and collecting the coefficient of ε^0, we obtain

$$\frac{d^2 y_0}{d\xi^2} - y_0 = 0, \quad y(\xi = 0) = 1$$

Solving this,

$$y_0(\xi) = be^{-\xi} = Y_i(\xi)$$

Using the matching technique, we obtain $b = 1$; thus, $Y_i(\xi) = e^{-\xi}$.

Matching:

$$\lim_{\xi \to \infty} e^{-\xi} = \lim_{x \to 0} 0 = 0$$

Composite solution: Hence,

$$y_{comp}(x) = Y_0(x) + Y_i(\xi) - \text{Match}$$

$$= 0 + e^{-\xi} - 0 = e^{-\xi} = e^{-x/\sqrt{\varepsilon}}$$

Note: The exact solution is also $y(x) = e^{-x/\sqrt{\varepsilon}}$ (Figure 4.7).

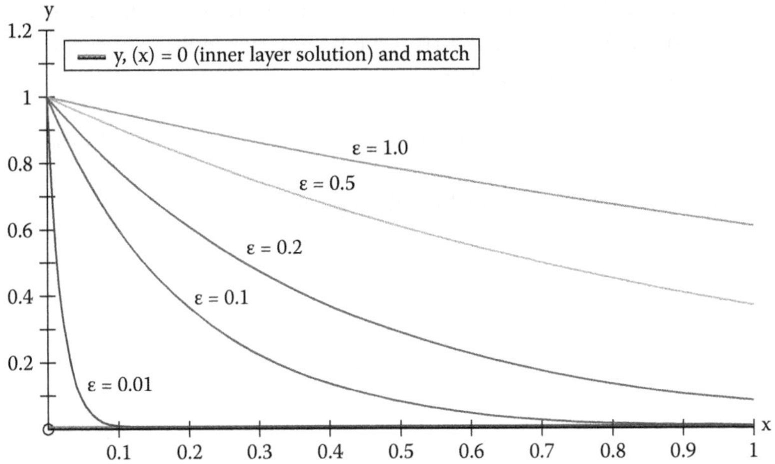

Figure 4.7 Plot of exact solutions of Example 4.10.

Example 4.11

Solve the differential equation

$$\varepsilon \frac{d^2 y}{dx^2} + 4\frac{dy}{dx} + 4y = 0, \quad y(0) = 0, \quad y(1) = 1 \text{ for } 0 < x < 1$$

Solution: Substitute

$$y(x) = y_0(x) + \varepsilon y_1(x) + \varepsilon^2 y_2(x) + \ldots$$

in the given differential equation; then, we obtain

$$\varepsilon\left(\frac{d^2 y_0}{dx^2} + \varepsilon\frac{d^2 y_1}{dx^2} + \ldots\right) + 4\left(\frac{dy_0}{dx} + \varepsilon\frac{dy_1}{dx} + \ldots\right) + 4\left(y_0 + \varepsilon y_1 + \ldots\right) = 0 \tag{4.13}$$

Coefficient of $\varepsilon^0 : \frac{dy_0}{dx} + y_0 = 0$

Its general solution is

$$y_0(x) = ce^{-x} \tag{4.14}$$

where c is an arbitrary constant. Clearly, this solution cannot satisfy both the boundary conditions simultaneously, but it can provide a solution by considering one boundary condition at a time. As we have two boundary conditions, we can find two solutions, one using $y(0) = 0$ and the other using $y(1) = 1$. Thus we have two cases.

Case 1: If we assume a boundary layer occurs at $x = 0$, that is, Equation (4.14) does not satisfy the boundary condition at $x = 0$, then the ***outer layer solution*** is

$$Y_o(x) = y_0(x) = e^{1-x} \tag{4.15}$$

Let us introduce the transformation given by

$$\xi = \frac{x}{\varepsilon^{\alpha}} \tag{4.16}$$

Using consistency check $\alpha = 1$ in the given differential equation, it becomes

$$\varepsilon^{1-2\alpha}Y'' + 4\varepsilon^{-\alpha}Y' + 4Y = 0, \quad Y(0) = 0$$

Here, the appropriate expansion within the boundary layer will be

$$Y(\xi) = Y_0(\xi) + \varepsilon Y_1(\xi) + \varepsilon^2 Y_2(\xi) + \ldots$$

substituting in Equation (4.13) and putting $\alpha = 1$, we have

$$\left(\frac{d^2Y_0}{d\xi^2} + \varepsilon\frac{d^2Y_1}{d\xi^2} + \varepsilon^2\frac{d^2Y_2}{d\xi^2} + \ldots\right) + 4\left(\frac{dY_0}{d\xi} + \varepsilon\frac{dY_1}{d\xi} + \varepsilon^2\frac{dY_2}{d\xi}\right)$$
$$+ 4\varepsilon\left(Y_0(\xi) + \varepsilon Y_1(\xi) + \varepsilon^2 Y_2(\xi) + \ldots\right) = 0$$

To solve for $Y_0, Y_1, Y_2, \ldots$ by balancing the terms in the equation by order of ε:

$$\varepsilon^0 : \frac{d^2Y_0}{d\xi^2} + 4\frac{dY_0}{d\xi} = 0, \quad Y_0(0) = 0 \tag{4.17}$$

The general solution of this is

$$Y_0(\xi) = c_1(1 - e^{-4\xi}) \tag{4.18}$$

where c_1 is the arbitrary constant.

Matching:

$$\lim_{\xi\to\infty} Y_0(\xi) = \lim_{x\to 0}(Y_o(x) = y_0(x))$$

$$\lim_{\xi\to\infty} c_1(1 - e^{-4\xi}) = \lim_{x\to 0} e^{1-x} \Rightarrow c_1 = e$$

Substituting the value of c_1 in Equation (4.18), we obtain

$$Y_0(\xi) = e(1 - e^{-4\xi}) = e - e^{1-4\xi}$$

Matching solution:

$$\lim_{\xi \to \infty} Y_0(\xi) = \lim_{x \to 0}(Y_o(x) = y_0(x))$$

$$\lim_{\xi \to \infty}(e - e^{1-4\xi}) = \lim_{x \to 0} e^{1-x} = e \quad \text{(overlapping part)}$$

Hence,

$$y_{comp}(x) \sim y_0(x) + Y_0(\xi) - Match\ (over\ lapping\ part)$$

and thus,

$$y_{comp}(x) \sim e^{1-x} + \left(e - e^{1-\frac{4x}{\varepsilon}}\right) - e \sim e^{1-x} - e^{1-\frac{4x}{\varepsilon}}$$

which is the required approximate first term of both inner- and outer-layer asymptotic approximate solution.

Note that the exact solution to the problem is

$$y(x) = \frac{e^{\frac{2(\sqrt{1-a}+1)(x-1)}{a}}\left(e^{\frac{4\sqrt{1-ax}}{a}} - 1\right)}{e^{\frac{4\sqrt{1-a}}{a}} - 1} \tag{4.19}$$

Case 2: If we assume boundary layer at $x = 1$, i.e., Equation (4.14), then the outer solution is

$$Y_o(x) = Y_0(x) = 0 \tag{4.20}$$

Let us introduce the transformation

$$\xi = \frac{x-1}{\varepsilon^{\alpha}} \tag{4.21}$$

Then the given equation with the constant becomes

$$\varepsilon^{1-2\alpha} Y'' + 4\varepsilon^{-\alpha} Y' + 4Y = 0, \quad Y(0) = 1 \tag{4.22}$$

again, using consistency check $\alpha = 1$.

Assuming

$$Y(\alpha) = Y_0(\xi) + \varepsilon\, Y_1(\xi) + \ldots$$

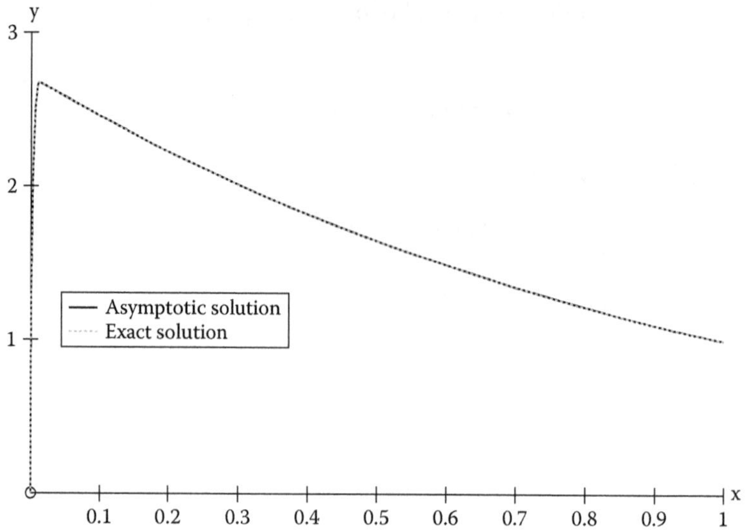

Figure 4.8 Plot of exact solution and asymptotic solutions of the equation.

and proceeding as in Case 1, we have

$$Y_0(\xi) = e^{-4\xi} + c_1(1 - e^{-4\xi}) \tag{4.23}$$

where c_1 is the arbitrary constant.

Matching:

$$\lim_{\xi \to \infty} Y_0(\xi) = \lim_{x \to 1}(Y_o(x) = y_0(x))$$

is not possible for any c_1.

Thus, there is no boundary layer at x = 1.

Figure 4.8 shows the plot of the asymptotic or composite solution and exact solutions for $\varepsilon = 0.01$. It can be observed that the first approximation of the asymptotic solution itself coincides with the exact solution.

Note: By using this procedure, we can also find the second term, and when added to the composite expansion presented, it gives a more accurate approximation.

Note: In Example 4.11 we tried finding out whether there are boundary layers at x = 0 and x = 1 and finally arrived at the conclusion saying that there is no second boundary layer.

We can avoid checking for another, known one, provided, we can guess the number of boundary layers beforehand. For this, examine the solution to the given differential equation by letting $\varepsilon = 0$. If the solution to that equation can satisfy one boundary condition (for a second order singularly perturbed equation), then there is one boundary layer in the domain of interest (refer to example mentioned above). But if the solution cannot satisfy both the boundary conditions, we can expect two boundary layers. The reader may verify the same for the boundary value problem given by $\varepsilon^2 \frac{d^2 y}{d x^2} + \varepsilon x \frac{d y}{d x} - y = -e^x$, $y(0) = 2$, $y(1) = 1$ for $0 < x < 1$.

Exercise Problems

1. Solve

$$y'' + \varepsilon y^2 = 0, \quad y(0) = 1, \quad y(1) = 0$$

2. Solve

$$y' + y = \varepsilon y^2, \quad y(0) = 1$$

3. Solve

$$\varepsilon y'' + y' + y = 0, \quad y(0) = 0, \quad y(1) = 1$$

4. Solve

$$\varepsilon y'' - y' + y = 0$$

   with $y(0) = 0$, $y(1) = 1$.

5. Solve

$$\varepsilon y' + y = 0, \quad y(0) = 1$$

6. Solve

$$\varepsilon y' = 1 - y, \quad y(0) = 0$$

7. Solve

$$y' + \varepsilon y^2 = 1, \quad y(0) = 0$$

Hints:

1.

$$y_0 = 1;\ y_1 = \frac{x(1-x)}{2};\quad y_2 = -\frac{x(1-x)\left(x^2-x-1\right)}{12}$$

and so on; the approximation is valid for all values of ε.

2.

$$y_0 = e^{-x},\quad y_1 = \left(e^{-x} - e^{-2x}\right),\ldots$$

The approximation is valid for all values of ε.

3.

$$Y_o(x) = e^{1-x},\quad Y_i(\xi) = e(1-e^{-\xi}),\quad \text{Match} = e$$

4.

$$y_{comp}(x) = e^{\frac{x-1}{\varepsilon}}$$

5.

$$y(x) = e^{\frac{-x}{\varepsilon}}$$

6.

$$y(x) = 1 - e^{\frac{-2x}{\varepsilon}}$$

7.

$$y_0 = x,\quad y_1 = \frac{-x^3}{3},\quad y_2 = \frac{-2x^5}{15},\ldots$$

The approximation is valid for all values of ε.

Applications

R. Bouyekhf and A. El Moudni. Regular perturbation for nonlinear discrete time systems: Application to electric circuit. *ACTA ET*, No. 1, 2004. Available at http://ie.utcluj.ro/files/acta/2004/Number%201/Paper05_Bouyekhf.pdf

F. Delhouse. Analytical treatment of air drag and earth oblateness effects upon an artificial satellite. *Celestial Mechanics and Dynamical Astronomy*, Vol. 52, pp. 85–103, 1991.

I.D. Feranchuk and L.I. Komarov. The regular perturbation theory in the strong-coupling polaron problem. *Journal of Physics C, Solid State Physics*, Vol. 15, No. 9, 1982.

I.D. Feranchuk and V.V. Riguk. Regular perturbation theory for two-electron atoms. *Physics Letters A*, Vol. 375, No. 26, pp. 2550–2554, 2011.

P.V. Kokotovic. Applications of singular perturbation techniques to control problems. *SIAM Review*, Vol. 26, No. 4, pp. 501–550, 1984.

J. Shinar. On applications of singular perturbation techniques in nonlinear optimal control. *Automata*, Vol. 19, No. 2, pp. 203–211, 1983.

G.P. Syrcos and P. Sannuti. Singular perturbation modeling and design techniques applied to jet engine control. *Optimal Control Applications and Methods*, Vol. 7, No. 1, pp. 1–17, 1986.

R.B. Washburn. Application of singular perturbation techniques (SPT) and continuation methods for on-line aircraft trajectory optimization. *IEEE Explore*, Vol. 17, pp. 983–990, 1978.

Y. Zou. An application of perturbation methods in evolutionary ecology. *Dynamics at the Horse Tooth*, Volume 2A, 2010.

Bibliography

R. Bellman. *Perturbation Techniques in Mathematics, Engineering and Physics.* New York: Dover, 1966.

C.M. Bender and S.A. Orszag. *Advanced Mathematical Methods for Scientists and Engineers.* Dordrecht, the Netherlands: Springer, 1999.

E.J. Hinch. *Perturbation Methods, Cambridge Texts in Applied Mathematics.* Cambridge, UK: Cambridge University Press, 1991.

M.H. Homes. *Introduction to Perturbation Methods.* Dordrecht, the Netherlands: Springer, 1995.

R.S. Johnson. *Singular Perturbation Theory: Mathematical and Analytical Techniques with Applications to Engineering.* Dordrecht, the Netherlands: Springer, 2005.

P. Kokotovic. *Singular Perturbation Methods in Control: Analysis and Design.* Philadelphia: SIAM, 1987.

A.H. Nayfeh. *Introduction to Perturbation Techniques.* New York: Wiley, 1993.

J.G. Simmonds. *A First Look at Perturbation Theory.* New York: Dover, 1986.

M. Van Dyke. *Perturbation Methods in Fluid Mechanics.* New York: Academic Press, 1964.

F. Verhulst. *Methods and Applications of Singular Perturbations: Boundary Layers and Multiple Timescale Dynamics.* Dordrecht, the Netherlands: Springer, 2005.

W. Wasow. *Asymptotic Expansions of Ordinary Differential Equations.* New York: Wiley, 1965.

5

Method of Multiple Scales

Introduction

The method of multiple scales, like the Lindstedt–Poincaré technique described in Chapter 3 of this book, helps us to construct uniformly valid approximate solutions to perturbation problems. The solutions thus obtained are valid for both small and large values of the independent variable.

In this method, we introduce a new set of variables called "fast-scale" and "slow-scale" variables in place of the independent variable and subsequently treat them as new independent variables. These new variables help us to remove the secular terms (refer to Chapter 3 for more details) in the solution of the problem. Another interesting aspect of this technique is that these new variables also help us to analyze the solution on different scales, which is particularly more useful in analyzing situations for which a phenomenon is expected to happen on different timescales. For instance, consider the problem of weakly damping oscillator described by the initial value problem (IVP)

$$u'' + \varepsilon u' + u = 0, \quad u(0) = 0, \quad u'(0) = 1. \tag{5.1}$$

The exact solution to this problem is

$$u(t,\varepsilon) = \frac{1}{\sqrt{1-\varepsilon^2/4}} e^{-\varepsilon t/2} \sin(t\sqrt{1-\varepsilon^2/4}). \tag{5.2}$$

Here, we see that the amplitude of oscillations and the frequency terms change only in higher-order terms of the expansion because

$$\frac{1}{\sqrt{1-\varepsilon^2/4}} \approx 1 + \frac{\varepsilon^2}{8}. \tag{5.3}$$

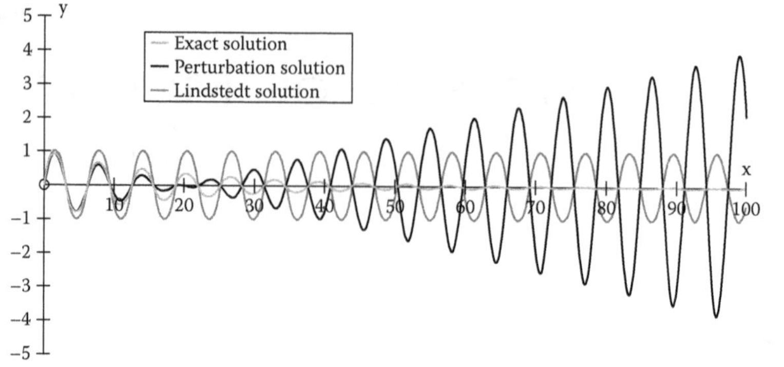

Figure 5.1 Plot of the exact, perturbation, and Lindstedt solutions of the weakly damping oscillator problem.

The perturbation solution to the problem in Equation (5.1) is

$$u(t,\varepsilon) = \sin t - \frac{\varepsilon}{2} t \sin t + \ldots \tag{5.4}$$

and the Lindstedt–Poincaré solution is given by

$$u(t,\varepsilon) = \sin t - \varepsilon \frac{\sin t \sin 2t}{4} + \ldots \tag{5.5}$$

Figure 5.1 shows the plot of the solutions given by Equations (5.2), (5.4), and (5.5). It is obvious from the plot that the perturbation solution (solid black) is a poor approximation to u when t is large as ε^{-1} in the second term is not small when compared with the first. In a way, the Lindstedt solution (dotted) is better, but then it failed to capture the actual behavior of the solution for large t.

Thus, we see that the effect (i.e., the damping effect) is insignificant on short timescales but predominant on long timescales.

Method of Multiple Scales

Consider the perturbation problem

$$L(u) + \varepsilon N(u) = 0 \tag{5.6}$$

where u is dependent on an independent variable t.

Let us introduce a new set of variables

$$T_0 = t; \quad T_1 = \varepsilon t; \quad T_2 = \varepsilon^2 t$$

and so on.

Here, T_0 is said to be a fast-scale variable, whereas T_1 is a slow-scale variable, T_2 is still a slower-scale variable, and so on. To illustrate this, let us take $\varepsilon = \frac{1}{60}$; then, if T_0 represents variations of the seconds arm, T_1 represents variations on the minutes hand, and T_2 represents the hours hand of a watch.

Now, we seek a solution to Equation (5.6) in the form

$$u(\varepsilon, T_0, T_1, T_2, \ldots) = u_0(T_0, T_1, T_2, \ldots) + \varepsilon u_1(T_0, T_1, T_2, \ldots) \quad (5.7)$$

Because

$$\frac{d}{dt} = \frac{\partial}{\partial T_0} + \varepsilon \frac{\partial}{\partial T_1} + \varepsilon^2 \frac{\partial}{\partial T_2} + \ldots$$

$$\frac{d^2}{dt^2} = \frac{\partial^2}{\partial T_0^2} + 2\varepsilon \frac{\partial^2}{\partial T_0 T_1} + \varepsilon^2 \left(\frac{\partial^2}{\partial T_1^2} + 2 \frac{\partial^2}{\partial T_0 T_2} \right) + \ldots$$

and so on.

Thus, Equation (5.6) now reduces to a system of partial differential equations that are to be solved for the unknown functions $u_0(T_0, T_1, T_2, \ldots), u_1(T_0, T_1, T_2, \ldots), \ldots$.

We illustrate this method next through some examples.

Example 5.1

Solve the equation

$$u'' + 2\varepsilon u' + u = 0, \quad u(0) = 1, \quad u'(0) = 0$$

where u' represents derivative with respect to t. Take two scale variables $T_0 = t; T_1 = \varepsilon t$.

Solution: Because

$$\frac{d}{dt} = \frac{\partial}{\partial T_0} + \varepsilon \frac{\partial}{\partial T_1}$$

and

$$\frac{d^2}{dt^2} = \frac{\partial^2}{\partial T_0^2} + 2\varepsilon \frac{\partial^2}{\partial T_0 T_1} + \varepsilon^2 \frac{\partial^2}{\partial T_1^2},$$

the given equation and the initial conditions take the form

$$\left(\frac{\partial^2}{\partial T_0^2}+2\varepsilon\frac{\partial^2}{\partial T_0 T_1}+\varepsilon^2\frac{\partial^2}{\partial T_1^2}\right)u(\varepsilon,T_0,T_1)$$

$$+2\varepsilon\left(\frac{\partial}{\partial T_0}+\varepsilon\frac{\partial}{\partial T_1}\right)u(\varepsilon,T_0,T_1)+u(\varepsilon,T_0,T_1)=0,$$

$$u(\varepsilon,0,0)=1,\left(\frac{\partial u}{\partial T_0}+\varepsilon\frac{\partial u}{\partial T_1}+\ldots\right)=0.$$

Let

$$u(\varepsilon,T_0,T_1)=u_0(T_0,T_1)+\varepsilon u_1(T_0,T_1)+\varepsilon^2 u_2(T_0,T_1)+\ldots.$$

Collecting the coefficient of $\varepsilon^0,\varepsilon^1,\ldots$, we obtain

$$\varepsilon^0:\begin{cases}\dfrac{\partial^2 u_0}{\partial T_0^2}+u_0=0,\\ u_0(0,0)=1,\\ \dfrac{\partial u_0}{\partial T_0}(0,0)=0\end{cases}\tag{5.8}$$

$$\varepsilon^1:\begin{cases}\dfrac{\partial^2 u_1}{\partial T_0^2}+u_1=-2\dfrac{\partial^2 u_0}{\partial T_0\,\partial T_1}-2\dfrac{\partial u_0}{\partial T_0},\\ u_1(0,0)=0,\\ \dfrac{\partial u_1}{\partial T_0}(0,0)=-\dfrac{\partial u_0}{\partial T_1}(0,0)\end{cases}\tag{5.9}$$

$$\varepsilon^2:\begin{cases}\dfrac{\partial^2 u_2}{\partial T_0^2}+u_2=-\dfrac{\partial^2 u_0}{\partial T_1^2}-2\dfrac{\partial u_0}{\partial T_1}-2\dfrac{\partial^2 u_1}{\partial T_0\,\partial T_1}-2\dfrac{\partial u_1}{\partial T_0},\\ u_2(0,0)=0,\\ \dfrac{\partial u_2}{\partial T_0}(0,0)=-\dfrac{\partial u_1}{\partial T_1}(0,0)\end{cases}\tag{5.10}$$

and so on.

Solving Equation (5.8), we have

$$u_0(T_0,T_1)=A(T_1)\cos T_0+B(T_1)\sin T_0 \text{ with } A(0)=1,B(0)=0.$$

Now, Equation (5.9) becomes

$$\begin{cases} \dfrac{\partial^2 u_1}{\partial T_0^2} + u_1 = 2\left(\dfrac{dA}{dT_1} + A\right)\sin T_0 - 2\left(\dfrac{dB}{dT_1} + B\right)\cos T_0, \\ u_1(0,0) = 0, \\ \dfrac{\partial u_1}{\partial T_0}(0,0) = 1 \end{cases} \tag{5.11}$$

Solving this partial differential equation, the complementary function of the general solution consists of $\sin T_0$ and $\cos T_0$. This gives rise to the so-called secular terms in view of the right-hand-side combination. Thus, to avoid secular terms in the solution, choose A and B such that

$$\frac{dA}{dT_1} + A = 0 \quad \text{with} \quad A(0) = 1$$

and

$$\frac{dB}{dT_1} + B = 0 \quad \text{with} \quad B(0) = 0$$

Thus, $A(T_1) = e^{-T_1}$ and $B(T_1) = 0$; hence, $u_0(T_0, T_1) = e^{-T_1}\cos T_0$.

Now, Equation (5.11) becomes

$$\begin{cases} \dfrac{\partial^2 u_1}{\partial T_0^2} + u_1 = 0, \\ u_1(0,0) = 0, \\ \dfrac{\partial u_1}{\partial T_0}(0,0) = 1 \end{cases}$$

whose solution is $u_1(T_0, T_1) = A^*(T_1)\cos T_0 + B^*(T_1)\sin T_0$ with $A^*(0) = 0, B^*(0) = 1$.

Consider Equation (5.10); substituting u_0 and u_1, we obtain

$$\begin{cases} \dfrac{\partial^2 u_2}{\partial T_0^2} + u_2 = e^{-T_1}\cos T_0 + 2\left(\dfrac{dA^*}{dT_1} + A^*\right)\sin T_0 - 2\left(\dfrac{dB^*}{dT_1} + B^*\right)\cos T_0, \\ u_2(0,0) = 0, \\ \dfrac{\partial u_2}{\partial T_0}(0,0) = 1 \end{cases}$$

Again, to eliminate the secular terms, as previously, we take

$$\frac{dA^*}{dT_1} + A^* = 0 \quad \text{with} \quad A^*(0) = 0$$

and

$$\frac{dB^*}{dT_1} + B^* = 0 \quad \text{with} \quad B^*(0) = 1$$

which gives $A^*(T_1) = 0$ and $B^*(T_1) = e^{-T_1}$; hence,

$$u_1(T_0, T_1) = e^{-T_1} \sin T_0.$$

Thus, the solution to the given problem is

$$u(T_0, T_1) = e^{-T_1} \cos T_0 + \varepsilon e^{-T_1} \sin T_0$$

or

$$u(t) = e^{-\varepsilon t} \cos t + \varepsilon e^{-\varepsilon t} \sin t$$

(up to the first approximation).

Figure 5.2 shows the plot of the exact solution

$$u(t) = e^{-\varepsilon t} \cos(\sqrt{1-\varepsilon^2}\, t),$$

the perturbation solution

$$u(t) = \cos t - \varepsilon t \cos t,$$

and the multiple-scale solution

$$u(t) = e^{-\varepsilon t} \cos t + \varepsilon e^{-\varepsilon t} \sin t.$$

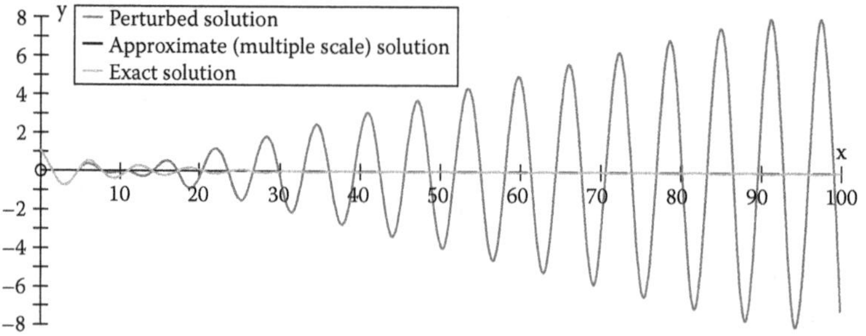

Figure 5.2 Plot of the exact, perturbation, and multiple-scale solutions.

Example 5.2

Solve the equation

$$u'' + u - \varepsilon u^3 = 0, \quad u(0) = 1, \quad u'(0) = 0$$

where u' represents derivative with respect to "t." Take two scale variables as $T_0 = t; T_1 = \varepsilon t$.

Solution: Because

$$\frac{d}{dt} = \frac{\partial}{\partial T_0} + \varepsilon \frac{\partial}{\partial T_1}$$

and

$$\frac{d^2}{dt^2} = \frac{\partial^2}{\partial T_0^2} + 2\varepsilon \frac{\partial^2}{\partial T_0 T_1} + \varepsilon^2 \frac{\partial^2}{\partial T_1^2},$$

the given equation takes the form

$$\left(\frac{\partial^2}{\partial T_0^2} + 2\varepsilon \frac{\partial^2}{\partial T_0 T_1} + \varepsilon^2 \frac{\partial^2}{\partial T_1^2} \right) u(\varepsilon, T_0, T_1) + u(\varepsilon, T_0, T_1) - \varepsilon u^3(\varepsilon, T_0, T_1) = 0,$$

$$u(\varepsilon, 0, 0) = 1, \left(\frac{\partial u}{\partial T_0} + \varepsilon \frac{\partial u}{\partial T_1} + \ldots \right)(\varepsilon, 0, 0) = 0.$$

Let

$$u(\varepsilon, T_0, T_1) = u_0(T_0, T_1) + \varepsilon u_1(T_0, T_1) + \varepsilon^2 u_2(T_0, T_1) + \ldots,$$

collecting the coefficient of

$$\varepsilon^0 : \begin{cases} \dfrac{\partial^2 u_0}{\partial T_0^2} + u_0 = 0, \\ u_0(0,0) = 1, \\ \dfrac{\partial u_0}{\partial T_0}(0,0) = 0 \end{cases} \tag{5.12}$$

$$\varepsilon^1 : \begin{cases} \dfrac{\partial^2 u_1}{\partial T_0^2} + u_1 = -2 \dfrac{\partial^2 u_0}{\partial T_0 \partial T_1} + u_0^3, \\ u_1(0,0) = 0, \\ \dfrac{\partial u_1}{\partial T_0}(0,0) = -\dfrac{\partial u_0}{\partial T_1}(0,0) \end{cases} \tag{5.13}$$

$$\varepsilon^2 : \begin{cases} \dfrac{\partial^2 u_2}{\partial T_0^2} + u_2 = -\dfrac{\partial^2 u_0}{\partial T_1^2} - 2\dfrac{\partial u_1}{\partial T_1 \partial T_0} + 3u_0^2 u_1, \\ u_2(0,0) = 0, \\ \dfrac{\partial u_2}{\partial T_0}(0,0) = -\dfrac{\partial u_1}{\partial T_1}(0,0) \end{cases} \tag{5.14}$$

and so on.

Solving Equation (5.12), we obtain

$$u_0(T_0, T_1) = A(T_1)e^{iT_0} + B(T_1)e^{-iT_0}$$

with

$$A(0) = \frac{1}{2}, B(0) = \frac{1}{2}$$

Note: We note that the solution to this equation also can be taken as

$$u_0(T_0, T_1) = A(T_1)\cos T_0 + B(T_1)\sin T_0,$$

but the present form is more convenient because of the nonlinear nature of the given equation.

Now, Equation (5.13) becomes

$$\begin{aligned} &\frac{\partial^2 u_1}{\partial T_0^2} + u_1 = -2\left(i\frac{dA}{dT_1}e^{iT_0} - i\frac{dB}{dT_1}e^{-iT_0} \right) + A^3 e^{3iT_0} + B^3 e^{-3iT_0} \\ &\qquad + 3A^2 B e^{iT_0} + 3AB^2 e^{-iT_0}, \\ &u_1(0,0) = 0, \\ &\frac{\partial u_1}{\partial T_0}(0,0) = 1 \end{aligned} \tag{5.15}$$

To avoid secular terms in the solution, choose A and B such that

$$-2i\frac{dA}{dT_1} + 3A^2 B = 0 \quad \text{with} \quad A(0) = \frac{1}{2}$$

and

$$2i\frac{dB}{dT_1} + 3AB^2 = 0 \quad \text{with} \quad B(0) = \frac{1}{2}$$

Thus,

$$A(T_1)=\frac{1}{2}e^{-\frac{3}{8}iT_1} \quad \text{and} \quad B(T_1)=\frac{1}{2}e^{\frac{3}{8}iT_1},$$

hence,

$$u_0(T_0,T_1)=\cos\left(T_0-\frac{3}{8}T_1\right).$$

Now, Equation (5.13) becomes

$$\frac{\partial^2 u_1}{\partial T_0^2}+u_1=\frac{1}{4}\cos\left(\frac{9}{8}T_1-3T_0\right),$$

$$u_1(0,0)=0,$$

$$\frac{\partial u_1}{\partial T_0}(0,0)=-\frac{\partial u_0}{\partial T_1}(0,0)=0$$

whose solution is

$$u_1(T_0,T_1)=-\frac{1}{32}\cos\left(\frac{9}{8}T_1-3T_0\right)$$

hence, a first approximate solution to the problem is

$$u(\varepsilon,T_0,T_1)=u_0(\varepsilon,T_0,T_1)+\varepsilon u_1(\varepsilon,T_0,T_1)$$

(up to the first approximation).

Remarks: You may note that the method of multiple scales has drawn the attention of several researchers. In the past, the method was reexamined almost once every 6 months, a process that continued for years. The technique described in this chapter was developed by Sturrock (1957), Frieman (1963), Nayfeh (1965a, 1965b), and Sandri (1965, 1967). The second version of the method of multiple scales was introduced and applied to various problems by Cole and Kevorkian (1963). Refer to the work of Nayfeh (1972) for more details on this method and also a list of applications of this method in varied fields like flight mechanics, solid mechanics, fluid mechanics, atmospheric science, and so on.

Exercise Problems

1. Solve

$$u'' + \varepsilon u' + u = 0, \quad u(0) = 0, \quad u'(0) = 1$$

using the multiple-scale technique.
Hint:

$$u_0(T_0, T_1) = e^{-T_1/2} \sin(T_0)$$

2.

$$u'' - \varepsilon((1 - u^3)u' + u^3) + 9u = 0, \quad u(0) = 1, \quad u'(0) = 0$$

Hint:

$$u_0(T_0, T_1) = e^{T_1/2} \cos\left(3T_0 - \frac{1}{8}\left(1 - e^{T_1}\right)\right)$$

Applications

H.R. Abbasi, A. Gholami, S.H. Fathi, and A. Abbasi. Using multiple scales method and chaos theory for detecting route to chaos in chaotic oscillations in voltage transformer with nonlinear core loss models. *International Review on Modelling and Simulations*, Vol. 4, No. 5, pp. 2195–2210, 2011.

M.J. Anderson and P.G. Vaidya. Application of the method of multiple scales to linear waveguide boundary value problems. *Journal of the Acoustics Society of America*, Vol. 88, p. 2450, 1990. http://dx.doi.org/10.1121/1.400085.

O.R. Asfar and A.H. Nayfeh. The application of the method of multiple scales to wave propagation in periodic structures. *SIAM Review*, Vol. 25, pp. 455–480, 1983.

M. Bataineh and O. Rafik Asfar. Application of multiple scales analysis and the fundamental matrix method to rugate filters: Initial-value and two-point boundary problem formulations. *Journal of Lightwave Technology*, Vol. 18, pp. 2217–2223, 2000.

H. Dai and L. Wang. Vortex-induced vibration of pipes conveying fluid using the method of multiple scales. *Theoretical and Applied Mechanics Letters*, Vol. 2, p. 022006, 2012.

M. Janowicz. Method of multiple scales in quantum optics. *Physics Reports*, Vol. 375, pp. 327–410, 2003.

A. Shooshtari and S.E. Khadem. A multiple scale method solution for the nonlinear vibration of rectangular plates. *Scientia Iranica*, Vol. 14, pp. 64–71, 2007.

Bibliography

C.M. Bender and S.A. Orzag. *Advanced Mathematical Methods for Scientists and Engineers*. New York: Springer, 1999.

J.D. Cole and J. Kevorkian. Uniformly valid asymptotic approximations for certain nonlinear differential equations. In *Nonlinear Differential Equations and Nonlinear Mechanics*. New York: Academic, 1963, pp.113–130.

E.A. Frieman. On a new method in the theory of irreversible processes. *Journal of Mathematical Physics*, Vol. 4, pp. 410–418, 1963.

M.H. Holmes. *Introduction to Perturbation Methods*. New York: Springer, 1995.

J. Kevorkian and J.D. Cole. *Multiple Scale and Singular Perturbation Methods*. New York: Springer, 1996.

A.H. Nayfeh. Nonlinear oscillations in a hot electron plasma. *Physics of Fluids*, Vol. 8, pp. 1896–1898, 1965a.

A.H. Nayfeh. A perturbation method for treating nonlinear oscillation problems. *Journal of Mathematics and Physics*, Vol. 44, pp. 368–374, 1965b.

A.H. Nayfeh. *Perturbation Methods*. New York: Wiley, 1972.

G. Sandri. A new method of expansion in mathematical physics. *Nuovo Cimento*, pp. 67–93, 1965.

G. Sandri. Uniformation of asymptotic expansions. In *Nonlinear Partial Differential Equations: A Symposium of Methods of Solutions*. New York: Academic, 1967, pp. 259–277.

P.S. Sturrock. Nonlinear effects in electron plasmas. *Proceedings of the Royal Society London A*, Vol. 242, No. 1230, pp. 277–299, 1957.

M. Van Dyke. *Perturbation Methods in Fluid Mechanics*. Stanford, CA: Parabolic Press, 1975.

6 WKB Theory

Introduction

In this chapter, we introduce another asymptotic method for solving singularly perturbed problems called the WKB approximation, named after the physicists Gregor Wentzel, Hendrik Kramers, and Léon Brillown, who all developed it in 1926. This approximation is considered to be the most powerful tool for obtaining global approximations to the solutions of singularly perturbed equations but is applicable only to linear differential equations. Although this method was developed basically for solving singularly perturbed problems, it can be used to solve unperturbed problems as well. Note also that the problems involving large parameters and the problems for which we have to find solutions at irregular singular points (IRSPs) also can be solved easily using this method (see Example 6.1 of this chapter). In addition, problems with rapidly oscillating solutions can be solved using this method (Bender and Orszag, 1999).

The WKB approximation particularly finds applications in problems related to quantum mechanics in which the wave propagation is such that the frequency of the wave is large or the wavelength of the wave is small. Another interesting observation is that the solution provided by this approximate method coincides with the exact solution for some equations, as can be seen in the examples in the following section.

The present chapter is arranged as follows: Section 6.2 presents the WKB method or approximation for unperturbed problems, followed by discussion of the method for singularly perturbed problems in Section 6.3.

WKB Approximation for Unperturbed Problems

Let us first see the WKB method for an unperturbed problem whose equation is of the form

$$\frac{d^2y}{dx^2}+f(x)y=0 \tag{6.1}$$

where $f(x)$ is a slowly varying function of x.

The solution of Equation (6.1) with $f(x)$ as a constant suggests that the solution can be taken in the form of

$$y(x)=e^{i\psi(x)} \tag{6.2}$$

Differentiating this equation with respect to x, we have

$$y'(x)=ie^{i\psi(x)}\psi'(x) \tag{6.3}$$

Differentiating this equation again, we obtain

$$y''(x)=i^2e^{i\psi(x)}\left(\psi'(x)\right)^2+ie^{i\psi(x)}\psi''(x) \tag{6.4}$$

Now, substituting Equations (6.4) and (6.2) in Equation (6.1), we obtain

$$-\psi'^2(x)+i\psi''(x)+f(x)=0 \tag{6.5}$$

Let us assume that $\psi''(x)$ is small; then, a first approximation using the Equation (6.5) can be obtained as

$$\psi'(x)=\pm\sqrt{f(x)} \tag{6.6}$$

$$\psi(x)=\pm\int\sqrt{f(x)}dx \tag{6.7}$$

The condition of validity of the above form is [that the $\psi''(x)$ be small] that (Shinn, 2010)

$$\left|\psi''(x)\right|\simeq\frac{1}{2}\left|\frac{f'(x)}{\sqrt{f(x)}}\right|\leq\left|f(x)\right| \tag{6.8}$$

From Equations (6.2) and (6.8), we see that $\frac{1}{\sqrt{f(x)}}$ is roughly $\frac{1}{2\pi}$ times one "wave length" or one "exponential length" of the solution $y(x)$. Thus, the condition of validity of our approximation is intuitively reasonable as the change in $f(x)$ in one wave length should be small when compared to $|f(x)|$ (Howison, 2005).

To find the second approximation, we proceed as follows: Substituting Equation (6.8) in Equation (6.5), we obtain

$$(\psi'(x))^2 \approx f(x) \pm \frac{i}{2}\frac{f'(x)}{\sqrt{f(x)}} \tag{6.9}$$

$$\psi'(x) \approx \pm\sqrt{f(x)} + \frac{i}{4}\frac{f'(x)}{f(x)} \tag{6.10}$$

$$\psi(x) \approx \pm\left(\int\sqrt{f(x)}dx + \frac{i}{4}\ln f(x)\right) \approx \pm\left(\int\sqrt{f(x)}dx + \ln(f(x))^{\frac{i}{4}}\right) \tag{6.11}$$

Now, substitute $\psi(x)$ from Equation (6.11) in Equation (6.2) to obtain the general solution of the given problem as

$$y(x) \approx \frac{1}{(f(x))^{\frac{1}{4}}}\left\{c_1 \exp\left[i\int\sqrt{f(x)}\,dx\right] + c_2 \exp\left[-i\int\sqrt{f(x)}\,dx\right]\right\} \tag{6.12}$$

where c_1 and c_2 are arbitrary constants.

Thus, we have found an approximation to the general solution of the Equation (6.1) in a region where the condition of validity given by Equation (6.8) holds.

Note: This method fails if $f(x)$ changes too rapidly or if $f(x)$ vanishes in the interval of interest. The reason that the latter condition is a more serious difficulty is because of the typical behavior of the solution at the turning point. In the region where $f(x) > 0$, the solution is oscillatory; that is, in the region where $f(x) < 0$ is of an exponential nature.

To illustrate this, take $f(x) = x$, then the curve vanishes at $x = 0$.

For $f(x) > 0$, say $f(x) = x = 4$, the solutions are $\sin(2x)$ or $\cos(2x)$, which are oscillatory in nature, whereas for $f(x) = x = -4$, the solutions are exponential, and matching these solutions at $x = 0$ is not possible.

The point at which the function vanishes is said to be a turning point. Langer transformation is used to solve the so-called turning point problem, which is beyond the scope of this book (Nayfeh, 1972).

Suppose $f(x)$ satisfies the condition of validity of Equation (6.8) in the regions to both the left and the right of x_0; then, the following cases arise:

Case 1: If $x \ll x_0$ *and* $f(x) < 0$, then the solution is

$$y(x) \approx \frac{1}{\sqrt[4]{(-f(x))}}\left\{\exp\left[\int_x^{x_0}\sqrt{-f(x)}\,dx\right]+\exp\left[-\int_x^{x_0}\sqrt{-f(x)}\,dx\right]\right\} \tag{6.13}$$

Case 2: If $x \gg x_0$ *and* $f(x) > 0$, then

$$y(x) \approx \frac{1}{\sqrt[4]{(f(x))}}\left\{\exp\left[i\int_{x_0}^{x}\sqrt{f(x)}\,dx\right]+\exp\left[-i\int_{x_0}^{x}\sqrt{f(x)}\,dx\right]\right\} \tag{6.14}$$

Let us apply the method described to solve a few equations.

Example 6.1

Find the WKB function associated with the equation

$$y'' + xy = 0$$

where $x \gg 0$.

Solution: Comparing the given problem with Equation (6.1), we have $f(x) = x$. Because $x \gg 0$, The WKB approximate solution using Equation (6.14) is

$$y(x) = (x)^{\frac{-1}{4}} \exp\left(\pm i\int_0^x (x)^{\frac{1}{2}}\,dx\right) \tag{6.15}$$

that is,

$$y(x) = (x)^{-\frac{1}{4}} \exp\left(\pm i\frac{2}{3}(x)^{\frac{3}{2}}\right)$$

> **Note:** If $f(x) = -x$, then we use the formula in Equation (6.13) to get the WKB approximation. Observe that the solution is the same as that obtained using the asymptotic method (Example 3.2 of Chapter 3). (From this example, it is evident that the WKB method is powerful in the sense that it provides a solution in a single step.)

In literature, the equation in Example 6.1 is termed Airy's equation, and its solutions are called Airy's functions. The solution of equations with a turning point often involves these Airy functions (Bender and Orszag, 1999).

We can also find solutions to eigenvalue problems considered in Chapter 3 using this WKB approximation. (You may check this.)

WKB Approximation for Perturbed Problems

As seen in the previous section, we observe that the WKB approximation to a solution of a differential equation is structured in a simplistic way. The appropriate solution might be some function with a superb quality or perception. If we see the WKB approximation, it contains, by the order of powers, exponentials pertaining to the elementary integrals of algebraic functions but still can effectively describe the solution function of the given equation. This is the highlight of the method that made it so popular.

This approximation is well suited for linear differential equations of any order and to initial value and eigenvalue problems, as previously mentioned. It is also useful in the evaluation of integrals occurring in the solution of a differential equation. An important aspect is that this WKB theory can be regarded as a more general theory for finding solutions of singularly perturbed problems as it contains the boundary-layer theory (discussed in Chapter 4) as a special case (Bender and Orszag, 1999).

Some Special Features of Solutions Near the Boundary Layer

In the discussion of the boundary-layer method in Chapter 4, we showed how to construct an approximate solution to a singularly perturbed differential equation. The basic idea was that this construction requires one to match the slowly varying outer solutions to the rapidly varying inner solutions. Note that the outer solution remains smooth if we allow the perturbation parameter ε to approach 0+. But, in this limit, the inner solution becomes discontinuous across the boundary layer as the thickness of the boundary layer tends to 0. In such a case, we say that the solution suffers a local breakdown at the boundary layer as $\varepsilon \to 0+$. This kind of behavior is called dissipative behavior.

Solutions of a few other singularly perturbed differential equations exhibit the so-called global breakdown, where it oscillates rapidly for small ε and has discontinuity when $\varepsilon \to 0+$. This nature of solutions is called dispersive nature.

We come across a few applications for which the solution exhibits either or both of these behaviors. One may note that the mathematical function (i.e., the exponential function) exhibits both of these properties for different arguments. If the exponent is real, it exhibits the dissipative nature, and if the exponent is imaginary, it exhibits the dispersive nature.

Thus, we construct the WKB solution in terms of an exponential function as follows:

$$y(x) \sim A(x)e^{\frac{S(x)}{\delta}}, \quad \delta \to 0+ \tag{6.16}$$

Here, the function $S(x)$ represents the "phase" of a wave and is assumed to be a nonconstant and slowly varying function in the breakdown region.

Now, the following cases arise:

1. $S(x)$ is real, in which case there is a boundary-layer thickness δ.
2. $S(x)$ is imaginary, then there is a region of rapid oscillation characterized by waves having wavelength order δ.
3. $S(x)$ is constant: The behavior of $y(x)$ is characterized by the slowly varying amplitude function $A(x)$.

Formal WKB Expansion

The exponential approximation in Equation (6.16) is not in a form that is more suitable for deriving asymptotic approximations because the amplitude and phase functions given by $A(x)$ and $S(x)$, respectively, depend implicitly on δ. Hence, it is convenient to expand $A(x)$ and $S(x)$ as series in powers of δ. Thus, the form of the solution is taken as

$$y(x) \sim \exp\left[\frac{1}{\delta}\sum_{n=0}^{\infty} \delta^n S_n(x)\right], \quad \delta \to 0 \tag{6.17}$$

Now, let us see how to solve the following singularly perturbed differential equation: Consider the equation of the form

$$\varepsilon^2 y'' = P(x)y, \quad P(x) \neq 0 \tag{6.18}$$

(in literature, this equation is called *Schrödinger*'s equation). Let us assume its solution is in the form given by Equation (6.17).

Differentiating it twice, we obtain

$$y'(x) \sim \frac{1}{\delta}\sum_{n=0}^{\infty} \delta^n S_n'(x) \exp\left[\frac{1}{\delta}\sum_{n=0}^{\infty} \delta^n S(x)\right], \quad \delta \to 0 \tag{6.19}$$

$$y''(x) \sim \left[\frac{1}{\delta^2}\left(\sum_{n=0}^{\infty} \delta^n S_n'(x)\right)^2 + \frac{1}{\delta}\sum_{n=0}^{\infty} \delta^n S_n''(x)\right] \exp\left(\frac{1}{\delta}\sum_{n=0}^{\infty} \delta^n S_n(x)\right), \quad \delta \to 0 \tag{6.20}$$

Now, let us substitute Equations (6.17) and (6.20) in Equation (6.18) to obtain

$$\varepsilon^2\left[\frac{1}{\delta^2}\left(S_0'(x) + \delta S_1'(x) + \delta^2 S_2'(x) + \ldots\right)^2 + \frac{1}{\delta}\left(S_0''(x) + \delta S_1''(x) + \delta^2 S_2''(x) + \ldots\right)\right] \exp\left(\frac{1}{\delta}\sum_{n=0}^{\infty} \delta^n S_n(x)\right)$$

$$= P(x) \exp\left(\frac{1}{\delta}\sum_{n=0}^{\infty} \delta^n S_n(x)\right)$$

which simplifies to

$$\varepsilon^2\left[\frac{1}{\delta^2}\left(S_0'(x)+\delta S_1'(x)+\delta^2 S_2'(x)+\ldots\right)^2 + \frac{1}{\delta}\left(S_0''(x)+\delta S_1''(x)+\delta^2 S_2''(x)+\ldots\right)\right]=P(x).$$

Put $\delta=\varepsilon$; then, we obtain

$$\left[\left(S_0'(x)+\varepsilon S_1'(x)+\varepsilon^2 S_2'(x)+\ldots\right)^2+\varepsilon\left(S_0''(x)+\varepsilon S_1''(x)+\varepsilon^2 S_2''(x)+\ldots\right)\right]=P(x)$$

Now, comparing the powers of ε, the unknown functions $S_0(x)$, $S_1(x), S_2(x), \ldots$ can be determined.

The equation for $S_0(x)$ is known as the *Eikonal* equation, which is

$$\varepsilon^0: \quad S_0'^2(x)=P(x)$$

and its solution is

$$S_0(x)=\pm\int_{x_0}^{x}\sqrt{P(t)}\,dt. \tag{6.21}$$

The equation for $S_1(x)$ is called the *transport* equation:

$$\varepsilon^1: \quad 2S_0'(x)S_1'(x)+S_0''(x)=0$$

and its solution is

$$S_1(x)=\frac{-1}{4}\ln P(x). \tag{6.22}$$

Higher-order terms can be obtained by collecting the coefficients of higher powers of ε as

$$O(\varepsilon^n): \quad 2S_0'(x)S_n'(x)+S_{n-1}''(x)+\sum_{i=1}^{n-1}S_i'(x)S_{n-i}'(x)=0,\ n\geq 2 \tag{6.23}$$

Thus, Equation (6.17) can be written as

$$y(x)\sim\exp\left[\frac{1}{\varepsilon}\left(S_0(x)+\varepsilon S_1(x)+\varepsilon^2 S_2(x)+\ldots\right)\right].$$

By substituting the values of $S_0(x)$ and $S_1(x)$ in this equation, we have

$$y(x) \sim c_1 P^{\frac{-1}{4}}(x)\exp\left[\frac{1}{\varepsilon}\int_a^x \sqrt{P(t)}\,dt\right] + c_2 P^{\frac{-1}{4}}(x)\exp\left[-\frac{1}{\varepsilon}\int_a^x \sqrt{P(t)}\,dt\right]$$
(6.24)

where c_1 and c_2 are the arbitrary constants that can be determined by the given initial or boundary conditions and a is the fixed integration point.

Also, we can determine the unknown functions $S_2(x), S_3(x), S_4(x), \ldots$ from Equation (6.23) to find a more accurate approximate solution of the WKB series.

It can be shown that

$$S_2(x) = \pm\int_a^x \left[\frac{P''}{8P^{\frac{3}{2}}} - \frac{5(P')^2}{32P^{\frac{5}{2}}}\right] dt \tag{6.25}$$

$$S_3(x) = -\frac{P''}{16P^2} + \frac{5P'^2}{64P^3} \tag{6.26}$$

and so on.

Note: Consider the equation

$$\varepsilon^2 \frac{d^2 y}{dx^2} + f(x)y = 0, \quad 0 < \varepsilon < 1 \tag{6.27}$$

Comparing this equation with the *Schrödinger* equation, we see that

$$P(x) = -f(x)$$

Hence, its solution, using Equation (6.24), is

$$y(x,\varepsilon) \sim (f(x))^{\frac{-1}{4}} \exp\left(\pm\frac{i}{\varepsilon}\int_{x_0}^x \sqrt{f(s)}\,ds\right) \tag{6.28}$$

Let us now consider Airy's equation.

Example 6.2

Solve

$$\frac{d^2Y}{dX^2} - XY = 0 \tag{6.29}$$

(where X is large) by introducing a small perturbation parameter.

Solution: Let

$$X = \frac{x}{\delta} \tag{6.30}$$

where δ is small (so that X is large). Then, we have

$$\delta^2 \frac{d^2y}{dx^2} - \frac{xy}{\delta} = 0 \tag{6.31}$$

$$\delta^3 \frac{d^2y}{dx^2} - xy = 0 \tag{6.32}$$

By comparing Equation (6.18) with Equation (6.32), we obtain $\varepsilon = \delta^{\frac{3}{2}}$, and $f(x) = -x$ and a one-term approximation for two solutions:

$$y(x,\delta) \sim (-x)^{-\frac{1}{4}} \exp\left(\pm \frac{2i}{3}\left(\frac{x}{\delta}\right)^{\frac{3}{2}}\right) \tag{6.33}$$

We must have $f(x) = -x > 0$, so we take the fourth root of $(-x)$. Figure 6.1 plots the WKB solutions of Airy's equations for a large independent variable. The WKB approximations for different values of δ are shown in Figure 6.2.

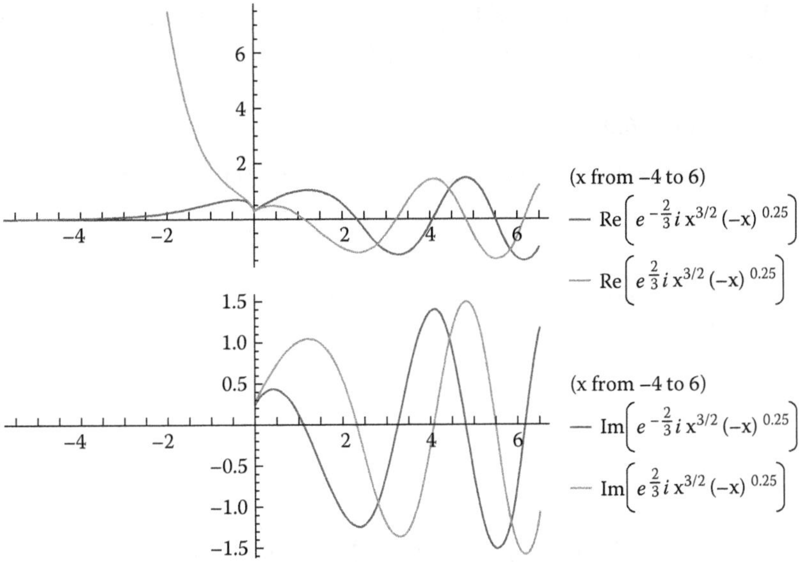

Figure 6.1 Plot of WKB solutions of Airy's equation for large independent variable.

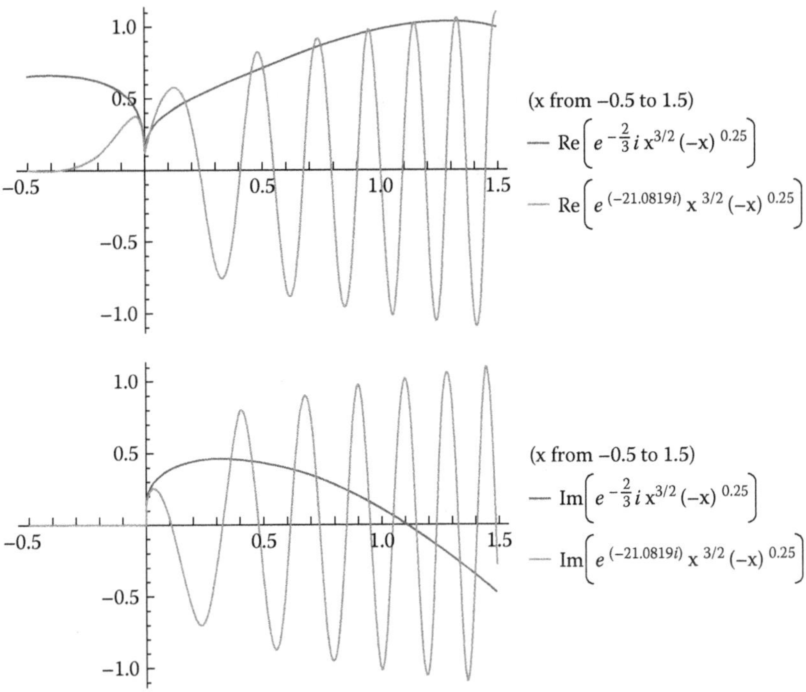

Figure 6.2 WKB approximation oscillatory solution for different values of δ.

We can also use the WKB approximation to solve a boundary value problem, as can be seen in Example 6.3.

Example 6.3

Using the WKB method, solve the boundary value problem

$$\varepsilon^2 y'' + 4y = 0, \quad y(0) = 0, \quad y(1) = 1$$

Solution: Comparing the given differential equation with *Schrödinger's* equation, Equation (6.18), we have

$$P(x) = -4 < 0$$

Thus,

$$S_0(x) = \pm \int^x \sqrt{P(t)}\,dt = \pm \int^x \sqrt{(-4)}dt = \pm i2x$$

and the WKB approximation is

$$y(x) = A\cos\left(\frac{2x}{\varepsilon}\right) + B\sin\left(\frac{2x}{\varepsilon}\right)$$

Using the boundary conditions, we obtain

$$A = 0,\ B = \frac{1}{\sin\left(\frac{2}{\varepsilon}\right)}$$

Hence, the required approximate solution is

$$y(x) = \frac{\sin\left(\frac{2x}{\varepsilon}\right)}{\sin\left(\frac{2}{\varepsilon}\right)} \tag{6.34}$$

Note that the exact solution to this problem is

$$y(x) = \frac{\sin\left(\frac{2x}{\varepsilon}\right)}{\sin\left(\frac{2}{\varepsilon}\right)}$$

where $\varepsilon \neq (n\pi)^{-1}$.

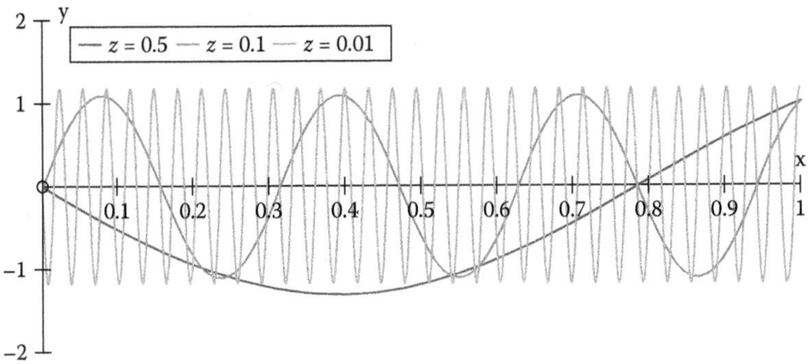

Figure 6.3 WKB approximation boundary-layer solution for different values of ε.

Note: Because $P(x) = -4 < 0$, $S_0(x)$ is purely imaginary, this accounts for the highly oscillatory nature of the solution, as seen in Figure 6.3.

Let us see one more example.

Example 6.4

Using the WKB method, solve the boundary-value problem

$$\varepsilon^2 y'' - 4y = 0, \quad y(0) = 0, \quad y(1) = 1$$

Solution: Comparing the given differential equation with *Schrödinger's* equation, Equation (6.18), we have

$$P(x) = 4 > 0$$

Thus,

$$S_0(x) = \pm \int^x \sqrt{P(t)}\,dt = \pm \int^x \sqrt{(4)}dt = \pm 2x$$

$$y(x) \sim \exp\left[\frac{\pm 2x}{\varepsilon}\right]$$

or

$$y(x) = A\cosh\left(\frac{2x}{\varepsilon}\right) + B\sinh\left(\frac{2x}{\varepsilon}\right)$$

Applying the boundary conditions, we obtain

$$A=0, \quad B=\frac{1}{\sinh\left(\frac{2}{\varepsilon}\right)}$$

Thus, the required approximate solution is

$$y(x)=\frac{\sinh\left(\frac{2x}{\varepsilon}\right)}{\sinh\left(\frac{2}{\varepsilon}\right)}$$

which is equivalent to the exact solution. Note that the exact solution is

$$y(x)=\frac{\sinh\left(\frac{2x}{\varepsilon}\right)}{\sin\left(\frac{2}{\varepsilon}\right)}$$

Note: Because $P(x)=4>0$, $S_0(x)$ is purely real, this corresponds to the occurrence of a boundary layer. We can observe it from the solution presented as well as from Figure 6.4.

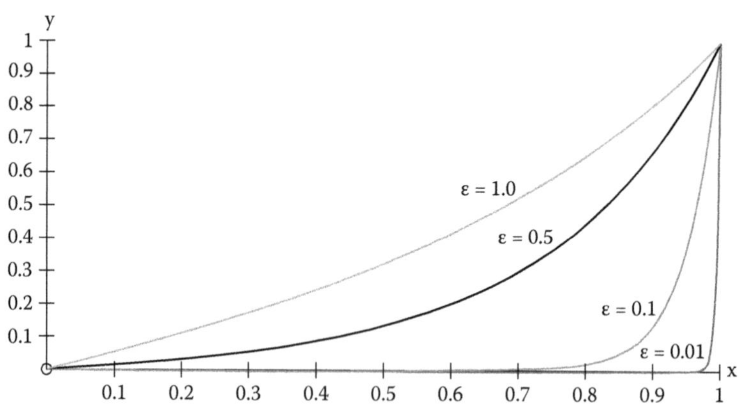

Figure 6.4 WKB approximation solution for different values of ε.

Now, let us show that the WKB approximation contains boundary-layer theory as a special case.

Example 6.5

Consider

$$\varepsilon y'' + a(x)y' + b(x)y = 0, \quad y(0) = A, \quad y(1) = B$$

Solution: Let

$$y(x) \sim \exp\left[\frac{1}{\delta}\sum_{n=0}^{\infty}\delta^n S_n(x)\right], \quad \delta \to 0$$

Then

$$y'(x) \sim \frac{1}{\delta}\sum_{n=0}^{\infty}\delta^n S_n'(x)\exp\left[\frac{1}{\delta}\sum_{n=0}^{\infty}\delta^n S(x)\right], \quad \delta \to 0 \qquad (6.35)$$

and

$$y''(x) \sim \left[\frac{1}{\delta^2}\left(\sum_{n=0}^{\infty}\delta^n S_n'(x)\right)^2 + \frac{1}{\delta}\sum_{n=0}^{\infty}\delta^n S_n''(x)\right]\exp\left(\frac{1}{\delta}\sum_{n=0}^{\infty}\delta^n S_n(x)\right), \quad \delta \to 0$$

Substituting these in the given differential equation, we obtain

$$\varepsilon\left[\frac{1}{\delta^2}\left(\sum_{n=0}^{\infty}\delta^n S_n'(x)\right)^2 + \frac{1}{\delta}\sum_{n=0}^{\infty}\delta^n S_n''(x)\right]\exp\left(\frac{1}{\delta}\sum_{n=0}^{\infty}\delta^n S_n(x)\right)$$
$$+ a(x)\frac{1}{\delta}\sum_{n=0}^{\infty}\delta^n S_n'(x)\exp\left(\frac{1}{\delta}\sum_{n=0}^{\infty}\delta^n S(x)\right)$$
$$+ b(x)\exp\left(\frac{1}{\delta}\sum_{n=0}^{\infty}\delta^n S_n(x)\right) = 0$$

which simplifies to

$$\varepsilon\left[\frac{1}{\delta^2}\left(\sum_{n=0}^{\infty}\delta^n S_n'(x)\right)^2 + \frac{1}{\delta}\sum_{n=0}^{\infty}\delta^n S_n''(x)\right] + a(x)\frac{1}{\delta}\sum_{n=0}^{\infty}\delta^n S_n'(x) + b(x) = 0$$

Expanding the summations, we have

$$\varepsilon\left[\frac{1}{\delta^2}\left(S_0'+\delta S_1'+\delta^2 S_2'+\ldots\right)^2+\frac{1}{\delta}\left(S_0''+\delta S_1''+\delta^2 S_2''+\ldots\right)\right]$$
$$+a(x)\frac{1}{\delta}\left(S_0'+\delta S_1'+\delta^2 S_2'+\ldots\right)+b(x)=0$$

or we have

$$\frac{\varepsilon}{\delta^2}S_0'^2+\varepsilon S_1'^2+\frac{2\varepsilon}{\delta}S_0'S_1'+\frac{\varepsilon}{\delta}S_0''+\varepsilon S_1''+a(x)\left(\frac{S_0'}{\varepsilon}+S_1'\right)+b(x)=0$$

Now, put $\delta=\varepsilon$ in this equation to obtain

$$\frac{1}{\varepsilon}S_0'^2+\varepsilon S_1^2+2S_0'S_1'+S_0''+\varepsilon S_1''+\frac{a(x)}{\varepsilon}S_0'+a(x)S_1'+b(x)=0$$

Collecting the coefficients of $\varepsilon^{-1},\varepsilon^0\ldots$

$$\varepsilon^{-1}: S_0'^2+a(x)S_0'=0 \tag{6.36}$$

$$\varepsilon^0: 2S_0'S_1'+S_0''+a(x)S_1'+b(x)=0 \tag{6.37}$$

From Equation (6.36), we obtain two solutions for S_0' as

$$S_0'=0 \quad \text{and} \quad S_0'=-a(x)$$

Case 1: When $S_0'=0$, then Equation (6.37) becomes

$$S_1'=\frac{-b(x)}{a(x)}$$
$$\Rightarrow S_0=c_1$$

and

$$S_1=-\int^x\frac{b(t)}{a(t)}dt$$

Hence, by WKB approximation, we obtain

$$y_1(x)\sim c_1\exp\left[-\int^x\frac{b(t)}{a(t)}dt\right],\ \varepsilon\to 0^+$$

Here, c_1 is an arbitrary constant that includes the term $e^{\frac{S_0}{S_0}}$.

This forms the outer solution of the boundary-layer problem considered.

Case 2: When

$$S_0' = -a(x)$$

then Equation (6.37) becomes

$$a(x)S_1' + a'(x) = b(x)$$

$$\Rightarrow S_0 = -\int^x a(t)\,dt$$

and

$$S_1 = -\ln a(x) + \int_0^x \frac{b(t)}{a(t)}\,dt$$

Hence, by WKB approximation we obtain

$$y_2(x) \sim c_2 \frac{1}{a(x)} \exp\left[\int_0^x \frac{b(t)}{a(t)}\,dt - \frac{1}{\varepsilon}\int_0^x a(t)\,dt\right], \quad \varepsilon \to 0^+$$

Here, c_2 is an arbitrary constant. This is the inner solution of the problem.

Hence, the general solution is

$$y(x) = y_1(x) + y_2(x)$$

which is

$$y(x) = c_1 \exp\left[-\int_0^x \frac{b(t)}{a(t)}\,dt\right] + c_2 \frac{1}{a(x)} \exp\left[\int_0^x \frac{b(t)}{a(t)}\,dt - \frac{1}{\varepsilon}\int_0^x a(t)\,dt\right], \quad \varepsilon \to 0^+ \tag{6.38}$$

Now, let us find the values of c_1 and c_2 using the given boundary conditions:

$$y(0) = A$$

gives

$$A = c_1 + \frac{c_2}{a(0)} \tag{6.39}$$

and

$$y(1) = B$$

gives

$$B = c_1 \exp\left[-\int_0^1 \frac{b(t)}{a(t)} dt\right] \Rightarrow c_1 = B \exp\left[\int_0^1 \frac{b(t)}{a(t)} dt\right] \qquad (6.40)$$

Neglecting the exponentially small term containing $\exp\left[-\varepsilon^{-1}\int_0^1 a(t)dt\right]$, we find that

$$A = B \exp\left[\int_0^1 \frac{b(t)}{a(t)} dt\right] + \frac{c_2}{a(0)} \Rightarrow c_2 = a(0)\left[A - B \exp\left[\int_0^1 \frac{b(t)}{a(t)} dt\right]\right]$$

$$y(x) \sim B \exp\left[\int_0^1 \frac{b(t)}{a(t)} dt\right] \exp\left[-\int_0^x \frac{b(t)}{a(t)} dt\right]$$

$$+ \frac{a(0)}{a(x)}\left[A - B \exp\left[\int_0^1 \frac{b(t)}{a(t)} dt\right]\right] \exp\left[\int_0^x \frac{b(t)}{a(t)} dt - \frac{1}{\varepsilon}\int_0^x a(t)dt\right], \quad \varepsilon \to 0^+$$

$$y(x) \sim B \exp\left[\int_x^1 \frac{b(t)}{a(t)} dt\right] + \frac{a(0)}{a(x)}\left[A - B \exp\left[\int_0^1 \frac{b(t)}{a(t)} dt\right]\right]$$

$$\exp\left[\int_0^x \frac{b(t)}{a(t)} dt - \frac{1}{\varepsilon}\int_0^x a(t)dt\right], \quad \varepsilon \to 0^+$$

Thus, the required boundary layer solution is

$$y(x) \sim B \exp\left[\int_x^1 \frac{b(t)}{a(t)} dt\right] + \left[A - B \exp\left[\int_0^1 \frac{b(t)}{a(t)} dt\right]\right] \exp\left[\frac{-a(0)x}{\varepsilon}\right] \qquad (6.41)$$

Let us now see the applications of this formula to find solutions of singularly perturbed problems.

Example 6.6

Solve the differential equation

$$\varepsilon \frac{d^2 y}{d x^2} + 4 \frac{d y}{d x} + 4 y = 0,$$

$$y(0) = 0, \quad y(1) = 1$$

for $0 < x < 1$.

Solution: By comparing the given equation with the general equation of Example 6.5, we have

$$a(x) = 4, \; b(x) = 4, A = 0 \text{ and } B = 1$$

Because $a(0) = 4$, we obtain, from Equation (6.41)

$$y(x) \sim \exp\left[\int_x^1 \frac{4}{4} dt\right] + \left[0 - \exp\left[\int_0^1 \frac{4}{4} dt\right]\right] \exp\left[\frac{-a(0)x}{\varepsilon}\right]$$

$$y(x) \sim \exp\left[\int_x^1 dt\right] + \left[0 - \exp\left[\int_0^1 dt\right]\right] \exp\left[\frac{-4x}{\varepsilon}\right]$$

$$y(x) \sim \exp[1 - x] - \exp\left[1 - \frac{4x}{\varepsilon}\right]$$

Hence, the required solution is

$$y(x) \sim e^{1-x} - e^{1-\frac{4x}{\varepsilon}}$$

Figure 6.4 shows the WKB approximation solution for different values of ε.

Example 6.7

Obtain an approximate solution to the boundary value problem

$$\varepsilon y'' + (1 + x) y' + y = 0, \quad y(0) = 1, \quad y(1) = 1$$

by the WKB approximation method.

Solution: By comparing the given equation with the general equation of Example 6.5, we have

$$a(x)=(1+x),\quad b(x)=1,\quad A=1$$

and

$$B=1$$

Here, $a(0)=1$; thus, the solution is

$$y(x)\sim \exp\left[\int_x^1 \frac{1}{(1+t)}dt\right]+\left[1-\exp\left[\int_0^1 \frac{1}{(1+t)}dt\right]\right]\exp\left[\frac{-x}{\varepsilon}\right]$$

which simplifies to

$$y(x)\sim \frac{2}{1+x}-e^{\frac{-x}{\varepsilon}}$$

Figures 6.5 and 6.6 show the WKB approximation solution for different values of ε.

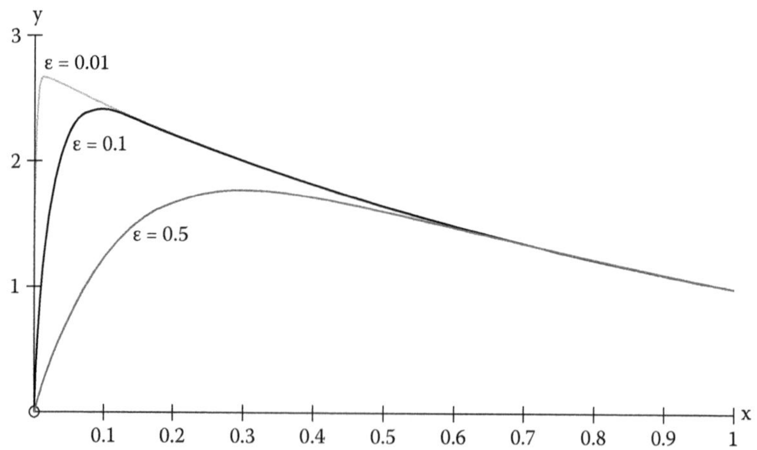

Figure 6.5 WKB approximation solution for different values of ε.

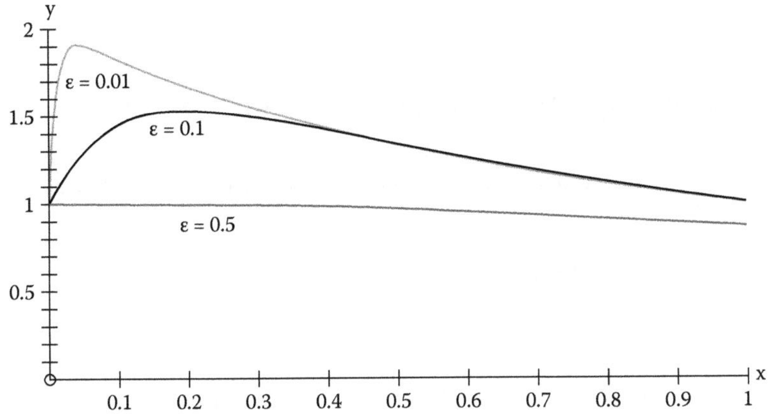

Figure 6.6 Plot of WKB solution for different values of ε.

Now, let us see how to solve the initial value problem (IVP) by the WKB method.

Example 6.8

Consider the IVP (Bender and Orszag, 1999)

$$\varepsilon^2 y'' = P(x)y, \quad y(0) = A, \quad y'(0) = B$$

Solution: The given differential equation is same as Example 6.3, and the solution of this problem using Equation (6.22) is

$$y(x) \sim c_1\left[P(x)\right]^{\frac{-1}{4}} \exp\left[\frac{1}{\varepsilon}\int_a^x \sqrt{P(t)}\,dt\right]$$

$$+ c_2\left[P(x)\right]^{\frac{-1}{4}} \exp\left[-\frac{1}{\varepsilon}\int_a^x \sqrt{P(t)}\,dt\right]$$

Let us take $a = 0$ (in view of the initial conditions) and differentiate this equation. Substituting in the given differential equation and using the initial conditions, we obtain

$$y(0) = (c_1 + c_2)\left[P(0)\right]^{\frac{-1}{4}} = A \qquad (6.42)$$

and

$$y'(0) = \frac{-1}{4}\left[P(0)\right]^{\frac{-5}{4}} P'(0)(c_1 + c_2) + \frac{\left[P(0)\right]^{\frac{1}{4}}}{\varepsilon}(c_1 - c_2) = B \quad (6.43)$$

If $A = 0$ and $B = 1$, then these equations can be written as follows:

Equation (6.42) becomes

$$c_1 + c_2 = 0 \Rightarrow c_1 = -c_2 \quad (6.44)$$

Equation (6.43) becomes

$$\frac{\left[P(0)\right]^{\frac{1}{4}}}{\varepsilon}(c_1 - c_2) = 1 \quad (6.45)$$

Solving Equations (6.44) and (6.45), we obtain

$$c_1 = \frac{\varepsilon}{2}\left[P(0)\right]^{\frac{1}{4}}$$

and

$$c_2 = \frac{-\varepsilon}{2}\left[P(0)\right]^{\frac{1}{4}}$$

Thus, the approximate solution to the given IVP is

$$y(x) \sim \varepsilon\left[P(x)P(0)\right]^{\frac{-1}{4}} \sinh\left[\int_0^x \sqrt{\frac{P(t)}{\varepsilon}}\, dt\right], \quad \varepsilon \to 0 \quad (6.46)$$

Example 6.9

Solve the IVP

$$\varepsilon^2 y'' = (1 - x^2)^2 y, \quad y(0) = 0, \quad y'(0) = 1$$

by the WKB method.

Solution: The solution of the given IVP can be obtained by substituting

$$P(x) = (1 - x^2)^2$$

and

$$P(0)=1$$

in Equation (6.46), and we obtain

$$y(x)\sim\frac{\varepsilon}{\sqrt{1-x^2}}\sinh\left[\frac{1}{\varepsilon}\left(x-\frac{x^3}{3}\right)\right],\quad \varepsilon\to 0^+$$

Note: Geometrical and Physical Optics

Geometrical Optics: If we consider only the first term of the WKB series, then the resultant approximation is known as geometrical optics and can be written as

$$y(x)\sim e^{\frac{S_0(x)}{\delta}}$$

We may note that this approximation does not constitute an asymptotic approximation to $y(x)$.

Physical Optics: If we consider the first two terms of the WKB series, then the resultant approximation is known as physical optics and can be written as

$$y(x)\sim e^{\frac{S_0(x)}{\delta}+S_1(x)},\quad \delta\to 0$$

Note: The approximation of physical optics expresses the leading asymptotic behavior of $y(x)$, while the approximation of geometrical optics contains the most rapidly varying component of the leading behavior.

In Section 6.3, we gave a few examples for geometrical optics. Now, let us examine the following example, which presents physical optics.

Example 6.10

Solve the given differential equation

$$\varepsilon y''+(1+x)y=0,\quad y(0)=0,\quad y(1)=1$$

Solution: Comparing the given differential equation with *Schrödinger's* equation, we have

$$P(x)=-(1+x)$$

then, using Equation (6.21), we obtain

$$S_0'(x) = \pm\sqrt{1+x}$$

$$\Rightarrow S_0(x) = \pm\frac{2i(1+x)^{\frac{3}{2}}}{3}$$

Hence, the solution of the geometrical optics yields

$$y(x) = \exp\left[\pm\frac{2i(1+x)^{\frac{2}{3}}}{3\varepsilon^{\frac{1}{2}}}\right]$$

or

$$y(x) = \frac{\sin\left[\frac{2i}{3\sqrt{\varepsilon}}(1+x)^{\frac{2}{3}}\right]}{\sin\left[\frac{2}{3\sqrt{\varepsilon}}2^{\frac{2}{3}}\right]}, \quad \varepsilon \neq \frac{\left(\frac{2}{3}\right)^2 2^{\frac{4}{3}}}{\left(n^2\pi^2\right)} \tag{6.47}$$

From Equation (6.22), we obtain

$$S_1(x) = \frac{-1}{4}\ln P(x)$$

Now, the physical optics can be written as

$$y(x) \sim e^{\frac{S_0(x)}{\varepsilon}+S_1(x)}, \quad \varepsilon \to 0$$

Hence, the general solution is written in the form

$$y(x) \sim c_1 P^{\frac{-1}{4}}(x)\exp\left[\frac{1}{\varepsilon}\int_a^x \sqrt{P(t)}\,dt\right] + c_2 P^{\frac{-1}{4}}(x)\exp\left[-\frac{1}{\varepsilon}\int_a^x \sqrt{P(t)}\,dt\right]$$

where

$$P(x) = -(1+x)$$

Hence, the solution of the physical optics is

$$y(x) = \frac{2^{\frac{1}{4}}\sin\left[\frac{2i}{3\sqrt{\varepsilon}}(1+x)^{\frac{2}{3}}\right]}{(1+x)^{\frac{1}{4}}\sin\left[\frac{2}{3\sqrt{\varepsilon}}2^{\frac{2}{3}}\right]}, \quad \varepsilon \neq \frac{\left(\frac{2}{3}\right)^2 2^{\frac{4}{3}}}{\left(n^2\pi^2\right)}$$

Exercise Problems

1. Find the WKB approximate solution to the boundary-value problem

$$\varepsilon^2 y'' - (1+x)^2 y = 0, \quad x > 0, \quad \varepsilon \ll 1, \quad y(0) = 1 \; and \; y(\infty) = 0$$

2. Find the WKB approximate solution to the boundary-value problem

$$\varepsilon y'' + \left(1 - \frac{x}{2}\right) y' - \frac{1}{2} y = 0, \quad x \in [0, 1], \quad y(0) = 0 \, and \; y(1) = 1$$

3. Find the WKB approximate solution of

$$\varepsilon y'' + y = 0, \quad y(0) = 0, \quad y(1) = 1 \, as \, \varepsilon \to 0^+$$

4. Find the physical and geometrical optics of the Airy function $y'' = xy$ in a *Schrödinger's* equation with $P(x) = x, \varepsilon = 1$.
5. Solve the equation

$$\varepsilon y'' - y = 0, \quad y(0) = 0, \quad y(1) = 1 \, as \, \varepsilon \to 0^+$$

by WKB approximation.

Hints:

1.

$$y(x) \sim \frac{1}{\sqrt{1+x}} e^{-\frac{1}{\varepsilon}\left(x + \frac{x^2}{2}\right)}$$

2.

$$y(x) = \frac{1}{2-x} - \frac{1}{2} e^{\frac{-x}{\varepsilon}}$$

3.

$$y(x) = \frac{\sin\left(\frac{x}{\sqrt{\varepsilon}}\right)}{\sin\left(\frac{1}{\sqrt{\varepsilon}}\right)}$$

4.

$$S_0 = \pm\frac{2}{3}x^{\frac{3}{2}}, S_1 = -\frac{1}{4}\ln x$$

5.

$$y(x) = \frac{\sinh\left(\frac{x}{\sqrt{\varepsilon}}\right)}{\sinh\left(\frac{1}{\sqrt{\varepsilon}}\right)}$$

Applications

Z.L. Gasyna and J.C. Light. Application of the WKB approximation in the solution of the Schrödinger equation. *Journal of Chemical Education*, Vol. 79, No. 1, p. 133, 2002.

Y. Matunobu. Application of the WKB method to the high subsonic flow of a compressible fluid past wedge profiles. *Journal of the Physical Society of Japan*, Vol. 17, No. 7, pp. 1181–1188, 1962.

R.G. Daghigh and M.D. Green. *Validity of the WKB Approximation in Calculating the Asymptotic Quasinormal Modes of Black Holes*. St. Paul, MN: Natural Sciences and Mathematics Departments, Metropolitan State University. arXiv: 1112.539v1 [gr-qc] 22-Dec 2011, pp. 1–10.

S.S. Wald and P. Lu. Application of the higher order modified WKB method to the Lennard–Jones potential. *Journal of Chemical Physics*, 61, p. 4680, 1974.

Bibliography

O.M. Bender and S.A. Orszag. *Advanced Mathematical Methods for Scientists and Engineers*. New York: McGraw-Hill, 1999.

S. Howison. *Practical Applied Mathematics—Modeling, Analysis and Approximation*. Cambridge, UK: Cambridge University Press, 2005.

A.H. Nayfeh. *Perturbation Methods*. New York: Wiley, 1972.

J. Shinn. Perturbation theory and WKB method. *Dynamics at Horse Show*, Vol. 2 (A), Focused Issue: Asymptotics and Perturbations, 2010.

7

Nonperturbation Methods

Introduction

This chapter continues to introduce methods to find approximate solutions to linear as well as nonlinear differential equations. In general, approximate analytical methods that we construct for any differential equation are based on two principles ("Basic Ideas and Brief History," n.d.):

(i) They have to provide analytic approximations efficiently.
(ii) They have to ensure that analytic approximations are accurate enough for all physical parameters.

It is well known that most of the perturbation solutions are valid only for small physical parameters. Thus, it is not guaranteed that a perturbation solution is valid in the whole range of all the values of physical parameters. Further, not every problem has a perturbation parameter involved in it, and even if one exists, the subproblems constructed may become so complicated that only a few of them have a solution. Thus, the perturbation techniques do not always satisfy each of the two principles.

To overcome these limitations of perturbation techniques, some traditional nonperturbation methods were developed. We consider initially two such methods in the present chapter:

1. Lyapunov's artificial parameter method; and
2. The δ expansion method.

Both of these methods are based on an artificial small parameter that will be introduced into the given problem, and the solution to the problem is taken in the form of a power series in terms of this artificial parameter. The advantage in these methods is that the artificial parameter can be chosen so that it is almost possible to find solutions to any given nonlinear problem. Thus, these methods in a way satisfy principle (i), but principle (ii) continues to fail, as illustrated in some of the examples shown in the subsequent sections.

We also introduce a novel method called the Adomian decomposition method (ADM), which is a powerful analytic technique for nonlinear differential equations. This is useful to solve ordinary and partial differential equations whether or not they contain small or large parameters. In this method, the given equation is split into linear and nonlinear parts (the nonlinear part may include some linear terms as well). Generally, the linear part is taken to be $\frac{d^k}{dx^k}$, where k is the highest-order derivative occurring in the given equation. As in the other methods, again the given problem is divided into subproblems whose solutions are obtained by integrating them k times with respect to x. Although this method satisfies principle (i), it cannot ensure the convergence of the approximation series and hence fails to satisfy principle (ii). Strictly in most of the cases, it is seen that the Adomian series converges quickly (Adomian, 1976; Adomian and Adomian, 1984). However, this is indeed great progress as these methods could provide solutions to a bigger set of problems.

The chapter is organized as follows: Section 7.2 deals with Lyapunov's artificial small-parameter method, Section 7.3 deals with the δ expansion method, and Section 7.4 deals with the ADM.

Lyapunov's Artificial Small-Parameter Method

In Lyapunov's artificial small-parameter method, we introduce an artificial parameter called the Lyapunov parameter into the given problem. The solution is assumed to be a power series in terms of this artificial parameter. As in the case of the power series solution method and the perturbation methods, here again we compare the coefficients of like powers of the artificial parameter. Thus, the given problem of solving a nonlinear ODE gives rise to linear subproblems whose solutions lead to a solution of the given problem.

This method is illustrated for a first-order differential equation: Consider the initial value problem (IVP) of the form

$$\frac{dy}{dx}+\left(y(x)\right)^n = k, \quad y(0)=b$$

where b and k are constants, and $n \geq 1$.

Let us introduce a small artificial parameter ε into the equation and rewrite it as

$$\frac{dy}{dx} + \varepsilon\left(y(x)\right)^n = k, \quad y(0) = b \tag{7.1}$$

The solution is taken to be a power series in terms of the artificial parameter ε. That is, we take

$$y(x) = y_0(x) + \varepsilon\, y_1(x) + \varepsilon^2 y_2(x) + \varepsilon^3 y_3(x) + \ldots \tag{7.2}$$

where $y_0(x), y_1(x), y_2(x)\ldots$ are unknown functions to be determined.

Differentiating $y(x)$ given in Equation (7.2) with respect to x, we have

$$\frac{dy}{dx} = \frac{dy_0}{dx} + \varepsilon\frac{dy_1}{dx} + \varepsilon^2\frac{dy_2}{dx} + \ldots \tag{7.3}$$

Substituting Equations (7.2) and (7.3) in Equation (7.1) and then comparing the like powers of ε on both sides, we obtain a system of linear differential equations. Solving these equations together with the given initial condition gives the values of $y_0(x), y_1(x), y_2(x), \ldots$.

Finally, substituting $y_0(x), y_1(x), y_2(x), \ldots$ in Equation (7.2), we obtain the desired solution $y(x)$ to the given IVP as

$$y(x) = \sum_{n=0}^{\infty} y_n(x)\varepsilon^n \tag{7.4}$$

We illustrate the method through specific examples.

Example 7.1

Solve the following IVP using Lyapunov's artificial parameter method:

$$\frac{dy}{dx} + y^3 = 3, \quad y(0) = 0$$

Solution: The equation is rewritten, after introducing Lyapunov's artificial parameter, as

$$\frac{dy}{dx} + \varepsilon\, y^3 = 3, \quad y(0) = 0 \tag{7.5}$$

Substituting Equations (7.2) and (7.3) in Equation (7.5), we have

$$\frac{dy_0}{dx}+\varepsilon\frac{dy_1}{dx}+\varepsilon^2\frac{dy_2}{dx}+\cdots+\varepsilon(y_0+\varepsilon y_1+\varepsilon^2 y_2+\ldots)^3=3,$$

$$y_0(0)+\varepsilon y_1(0)+\varepsilon^2 y_2(0)+\ldots=0 \tag{7.6}$$

Now, collecting the coefficients of like powers of ε from Equation (7.6), we obtain a system of linear equations given by

$$\varepsilon^0: \frac{dy_0}{dx}=3, \quad y_0(0)=0$$

$$\varepsilon^1: \frac{dy_1}{dx}+y_0^3=0, \quad y_1(0)=0$$

$$\varepsilon^2: \frac{dy_2}{dx}+3y_0^2 y_1=0, \quad y_2(0)=0$$

and so on.

Solving these equations together with the given conditions, we obtain

$$y_0=3x, \quad y_1=-\frac{27}{4}x^4, \quad y_2=\frac{729}{28}x^7,\ldots \tag{7.7}$$

Hence, the solution [using Equation (7.2)] is

$$y(x)=3x-\varepsilon\frac{27}{4}x^4+\varepsilon^2\frac{729}{28}x^7-\varepsilon^3\frac{729}{560}x^9(54x+35)+\ldots \tag{7.8}$$

In particular, if $\varepsilon=1$, the solution is

$$y(x)=3x-\frac{27}{4}x^4+\frac{729}{28}x^7-\frac{729}{560}x^9(54x+35)+\ldots$$

Example 7.2

Solve

$$\frac{d\theta}{dx}+\left(1+\frac{1}{x}\right)\theta=1, \quad \theta(0)=0$$

Introducing the artificial parameter ε, we have

$$\frac{d\theta}{dx}+\varepsilon\left(1+\frac{1}{x}\right)\theta=1, \quad \theta(0)=0 \tag{7.9}$$

Solution: Let

$$\theta = \theta_0 + \varepsilon\theta_1 + \varepsilon^2\theta_2 + \ldots$$

Computing $\frac{d\theta}{dx}$ and substituting in this above equation, we obtain

$$\frac{d\theta_0}{dx} + \varepsilon\frac{d\theta_1}{dx} + \varepsilon^2\frac{d\theta_2}{dx} + \cdots + \varepsilon\left(1 + \frac{1}{x}\right)(\theta_0 + \varepsilon\theta_1 + \varepsilon^2\theta_2 + \ldots) = 1$$

Comparing the coefficients of like powers of ε from the second equation, we have

$$\varepsilon^0 : \frac{d\theta_0}{dx} = 1, \quad \theta_0(0) = 0$$

$$\varepsilon^1 : \frac{d\theta_1}{dx} + \left(1 + \frac{1}{x}\right)\theta_0 = 0, \quad \theta_1(0) = 0$$

$$\varepsilon^2 : \frac{d\theta_2}{dx} + \left(1 + \frac{1}{x}\right)\theta_1 = 0, \quad \theta_2(0) = 0$$

and so on.

Solving these subproblems, we get

$$\theta_0 = x$$

$$\theta_1 = -x - \frac{x^2}{2}$$

$$\theta_2 = \frac{1}{12}\,(12\,\mathrm{x} + 9\,\mathrm{x}^2 + 2\,\mathrm{x}^3)$$

and so on.

Thus, an approximate solution using this problem is

$$\theta = x + \varepsilon\left(-x - \frac{x^2}{2}\right) + \varepsilon^2\left(x + \frac{3x^2}{4} + \frac{x^3}{6}\right) + \ldots \tag{7.10}$$

Note that the exact analytical solution to the problem is

$$\theta = 1 - \frac{1 - e^{-x}}{x}.$$

Figure 7.1 offers a comparison between the exact and the Lyapunov approximate solution.

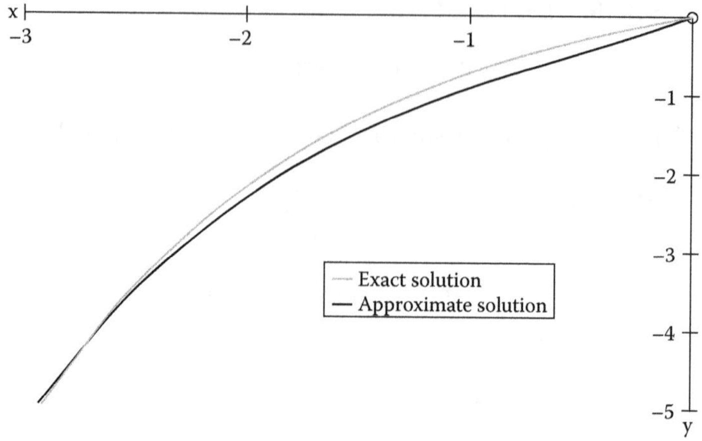

Figure 7.1 Comparison between the exact and the Lyapunov approximate solution.

Example 7.3

Solve

$$\frac{dy}{dx} + y^2 = 4, \quad y(0) = 0 \tag{7.11}$$

Solution: The equation is rewritten, after introducing Lyapunov's artificial parameter ε, as

$$\frac{dy}{dx} + \varepsilon y^2 = 4, \quad y(0) = 0 \tag{7.12}$$

Substituting Equations (7.2) and (7.3) in Equation (7.12), we obtain

$$\begin{gathered}\frac{dy_0}{dx} + \varepsilon \frac{dy_1}{dx} + \varepsilon^2 \frac{dy_2}{dx} + \cdots + \varepsilon(y_0 + \varepsilon y_1 + \varepsilon^2 y_2 + \ldots)^2 = 4, \\ y_0(0) + \varepsilon y_1(0) + \varepsilon^2 y_2(0) + \ldots = 0\end{gathered} \tag{7.13}$$

Now, collecting the coefficients of like powers of ε from Equation (7.13), we obtain a system of linear equations given by

$$\varepsilon^0 : \frac{dy_0}{dx} = 4, \quad y_0(0) = 0$$

$$\varepsilon^1 : \frac{dy_1}{dx} + y_0^2 = 0, \quad y_1(0) = 0$$

$$\varepsilon^2 : \frac{dy_2}{dx} + 2y_0 y_1 = 0, \quad y_2(0) = 0$$

$$\varepsilon^3 : \frac{dy_3}{dx} + 2y_0 y_1 + y_1^2 = 0, \quad y_3(0) = 0$$

and so on.

Solving these, we obtain

$$y_0 = 4x, \quad y_1 = -\frac{16}{3}x^3, \quad y_2 = \frac{128}{15}x^5, \quad y_3 = \frac{4352}{315}x^7, \ldots \quad (7.14)$$

Hence, the solution is

$$y(x) = 4x - \varepsilon\frac{16}{3}x^3 + \varepsilon^2\frac{128}{15}x^5 - \varepsilon^3\frac{4352}{315}x^7 + \ldots \quad (7.15)$$

Here, $y'(0) = 4$ from both Equations (7.12) and (7.15). Hence, this solution is valid for all values of ε. In particular, if $\varepsilon = 1$, the solution is

$$y(x) = 4x - \frac{16}{3}x^3 + \frac{128}{15}x^5 - \frac{4352}{315}x^7 + \ldots \quad (7.16)$$

Note: The exact solution of Equation (7.12) is

$$y(x) = \frac{2(e^{4x} - 1)}{(e^{4x} + 1)}.$$

For $\varepsilon = 1$, the approximate solution matches the exact solution for $0 < x \leq 0.6$, as can be seen from Figure 7.2.

Now, we illustrate the method for a coupled system.

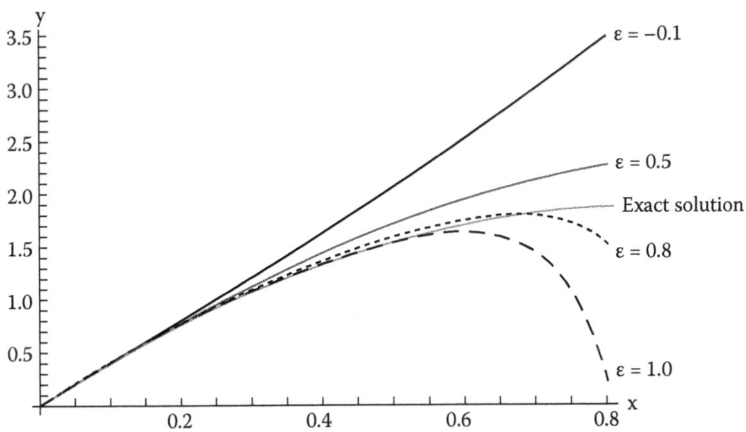

Figure 7.2 Comparison of the exact solution and approximate solution of the 20th order for various ε.

Example 7.4

Solve

$$f'' = \theta'$$

$$\theta'' + \frac{f\theta'}{2} = 0, \quad \theta(0) = 1, \quad \theta(1) = 0 \;\&\; f(0) = 0, \quad f'(1) = 1.$$

Solution: Rewriting these equations, we obtain

$$f'' - \varepsilon\theta' = 0 \tag{7.17}$$

$$\theta'' + \frac{\varepsilon f\theta'}{2} = 0 \tag{7.18}$$

Here,

$$f = f_0 + \varepsilon f_1 + \varepsilon^2 f_2 + \ldots$$

and

$$\theta = \theta_0 + \varepsilon\theta_1 + \varepsilon^2\theta_2 + \ldots \tag{7.19}$$

Now, Equations (7.17) and (7.18) can be written as

$$(f_0'' + \varepsilon f_1'' + \varepsilon^2 f_2'' + \ldots) - \varepsilon(\theta_0' + \varepsilon\theta_1' + \varepsilon^2\theta_2 + \ldots) = 0 \tag{7.20}$$

$$(\theta_0'' + \varepsilon\theta_1'' + \varepsilon^2\theta_2'' + \ldots) + \frac{\varepsilon}{2}\left(f_0 + \varepsilon f_1 + \varepsilon^2 f_2 + \ldots\right)\left(\theta_0' + \varepsilon\theta_1' + \varepsilon^2\theta_2' + \ldots\right) = 0 \tag{7.21}$$

Comparing the coefficients of like powers of ε on both sides of Equations (7.20) and (7.21), we have

$$\varepsilon^0 : f_0'' = 0 \;; \qquad \varepsilon^0 : \theta_0'' = 0$$

$$\varepsilon^1 : f_1'' - \theta_0' = 0 \;; \qquad \varepsilon^1 : \theta_1'' + \frac{f_0\theta_0'}{2} = 0$$

$$\varepsilon^2 : f_2'' - \theta_1' = 0 \;; \qquad \varepsilon^2 : \theta_2'' + \frac{f_0\theta_0' + f_0\theta_1' + f_1\theta_0'}{2} = 0$$

and so on.

From these equations, we obtain

$$f_0 = x \tag{7.22}$$

$$\theta_0 = -x + 1 \tag{7.23}$$

$$f_1 = x - \frac{x^2}{2} \tag{7.24}$$

$$\theta_1 = \frac{x^3}{12} - \frac{x}{12} \tag{7.25}$$

$$f_2 = -\frac{x^2}{24} + \frac{x^4}{48} \tag{7.26}$$

$$\theta_2 = -\frac{13x}{90} + \frac{25x^3}{144} - \frac{x^4}{48} - \frac{x^5}{120} \tag{7.27}$$

Substituting Equations (7.22), (7.24), and (7.26) in Equation (7.19) yields the approximate solution of f:

$$f = x + \varepsilon\left(x - \frac{x^2}{2}\right) + \varepsilon^2\left(\frac{-x^2}{24} + \frac{x^4}{48}\right) + \ldots \tag{7.28}$$

Similarly, by substituting Equations (7.23), (7.25), and (7.27) in Equation (7.19), we obtain the approximate solution of θ as

$$\theta = (1-x) + \varepsilon\left(\frac{-x}{12} + \frac{x^3}{12}\right) + \varepsilon^2\left(\frac{-13x}{90} + \frac{25x^3}{144} - \frac{x^4}{48} - \frac{x^5}{120}\right) + \ldots \tag{7.29}$$

Limitations: Although this method can be used for finding approximate analytical solutions for any given nonlinear equation, in theory we can place the artificial small parameter many different ways. Unfortunately, there are no theories to guide us regarding how to put it in a suitable place to obtain a better approximation.

Delta (δ) Expansion Method

Consider the algebraic equation $y^6 + y = 1$. This equation can be solved by the δ expansion method by rewriting it as $y^{1+\delta} + y = 1$.

Write $y^{1+\delta} = e^{(1+\delta)\ln y}$ and expand it in powers of δ. Now, compare the coefficients of different powers of δ; then, we arrive at a system of equations that are relatively easy to solve, and the solution to the equation can be obtained. Finally, put $\delta = 5$ (in view of the given algebraic equation) to obtain the solution of the algebraic equation considered.

Now, let us apply this logic for finding the solution of a differential equation.

Example 7.5

Solve the following nonlinear IVP by the δ expansion method.

$$\frac{dy}{dx}+y^3=3, \quad y(0)=0$$

Solution: This equation is written as

$$\frac{dy}{dx}+y^{1+\delta}=3 \tag{7.30}$$

Let

$$y(x)=y_0+\delta y_1+\delta^2 y_2+\ldots \tag{7.31}$$

Now,

$$y^{1+\delta}=y_0+(y_1+y_0 Log[y_0])\delta+(y_1+y_2+y_1 Log[y_0] + \frac{1}{2}y_0 Log[y_0]^2)\delta^2+\ldots \tag{7.32}$$

Equation (7.31) can be written as

$$\frac{dy}{dx}=\frac{dy_0}{dx}+\delta\frac{dy_1}{dx}+\delta^2\frac{dy_2}{dx}+\ldots \tag{7.33}$$

Substituting Equations (7.32) and (7.33) in Equation (7.30), we obtain

$$\left(\frac{dy_0}{dx}+\delta\frac{dy_1}{dx}+\delta^2\frac{dy_2}{dx}+\ldots\right) + y_0+(y_1+y_0 Log[y_0])\delta+(y_1+y_2+y_1 Log[y_0] + \frac{1}{2}y_0 Log[y_0]^2)\delta^2+\ldots=3 \tag{7.34}$$

Comparing the coefficients of different powers of δ of both sides of Equation (7.34), we have

$$\frac{dy_0}{dx}+y_0=3, \quad y_0(0)=0$$

$$\frac{dy_1}{dx}+y_1=-y_0 Log[y_0], \quad y_1(0)=0$$

$$\frac{dy_2}{dx}+y_2=-y_1\left(1+Log[y_0]-\frac{y_0}{2}Log[y_0]^2\right), \quad y_2(0)=0 \tag{7.35}$$

Solving Equation (7.35) using Mathematica®,

$$y_0 = 3 - e^{-x}$$

$$y_1 = \frac{-e^{-x}}{2}\begin{pmatrix} 6i\pi - \pi^2 - 3x^2 - 6Log[3] + 6e^{x}Log[3-3e^{-x}] \\ -6xLog[3-3e^{-x}] - 6Log[1-e^{-x}] \\ +6xLog[1-e^{-x}] + 6PolyLog[2, e^{x}] \end{pmatrix} \quad (7.36)$$

Putting $\delta = 2$ in Equation (7.36), we see that the approximate solution is

$$y = y_0 + 2y_1 + 4y_2 + \ldots \quad (7.37)$$

Adomian Decomposition Method

Introduction

Although now we are in a position to solve a few more nonlinear ordinary differential equations (ODEs) using the two methods described in the previous sections, the quest for yet a new method continued owing to the limitations of these methods. Hence, the ADM developed by George Adomian has emerged as an alternate method for solving a wide class of problems whose mathematical models involved algebraic, differential, integral, and integrodifferential equations; and higher-order ODEs and partial differential equations. The method is described next.

Adomian Decomposition Method

Let the general form of a nonlinear differential equation be as follows:

$$Fy = f \quad (7.38)$$

where F is a nonlinear differential operator, and y and f are functions of x.

Rewriting the equation in the form

$$Ly + Ry + Ny = f \quad (7.39)$$

where L is an easily invertible operator representing a part or whole of the linear portion of F, R is the remainder of the linear operator of F, and N is the nonlinear part of F.

Applying the inverse operator L^{-1} to Equation (7.39), the equation becomes

$$L^{-1}Ly = L^{-1}f - L^{-1}Ry - L^{-1}Ny \tag{7.40}$$

For instance, if L is a second-order differential operator, then L^{-1} is a twofold integration operator given by

$$L^{-1} = \iint (.)\, dx_1 dx_2 \tag{7.41}$$

Here, L^{-1} represents an integral operator.

$$L^{-1}(Ly) = \int_0^x\int_0^x (Ly(x))dxdx$$

$$= \int_0^x\int_0^x y''(x)dxdx$$

Using the given initial conditions, we obtain

$$L^{-1}Ly = y(x) - y(0) - xy'(0) \tag{7.42}$$

Then, Equation (7.40) becomes

$$y(x) - y(0) - xy'(0) = L^{-1}(f) - L^{-1}(Ry) - L^{-1}(Ny) \tag{7.43}$$

or

$$y(x) = \underbrace{y(0) + xy'(0) + L^{-1}(f)}_{y_0} - L^{-1}(Ry) - L^{-1}(Ny) \tag{7.44}$$

Hence, $y(x)$ can be written as

$$y(x) = y_0 - [L^{-1}(Ry) + L^{-1}(Ny)] \tag{7.45}$$

Let $y(x)$ be represented as a series given by (Adomian, 1992; "Adomian Decomposition Method," 2014)

$$y(x)=\sum_{n=0}^{\infty} y_n \tag{7.46}$$

and the nonlinear term $N(y)$ is written as an infinite series involving the Adomian polynomials $A_0, A_1, \ldots$ as follows:

$$N(y)=\sum_{n=0}^{\infty} A_n \tag{7.47}$$

where A_n's are obtained using the formula

$$A_n = \frac{1}{n!}\frac{d^n}{d\lambda^n}\left[N\left(\sum_{n=0}^{\infty}\lambda^n y_n\right)\right]_{\lambda=0} \tag{7.48}$$

Now, substituting Equations (7.46) and (7.47) in Equation (7.45), we obtain

$$\sum_{n=0}^{\infty} y_n = y_0 - L^{-1}\left[R\left(\sum_{n=0}^{\infty} y_n\right)\right] - L^{-1}\left(\sum_{n=0}^{\infty} A_n\right) \tag{7.49}$$

Consequently, Equation (7.49) can be written as

$$\begin{aligned}
y_0 &= y(0)+xy'(0)+L^{-1}(f)\\
y_1 &= -L^{-1}R(y_0)-L^{-1}(A_0)\\
&\ldots\ldots\ldots\ldots\ldots\ldots\\
y_{n+1} &= -L^{-1}R(y_n)-L^{-1}(A_n)
\end{aligned} \tag{7.50}$$

Solving this system gives

$$y_0(x),\quad y_1(x),\quad y_2(x),\ldots$$

and the solution is given by

$$y(x)=\sum_{k=0}^{n} y_k(x),\quad n>0 \tag{7.51}$$

Note: If $N(y)$ is a given nonlinear operator, then the Adomian polynomials can be calculated as follows (Hasan and Zhu, 2008):

$$\begin{aligned}
A_0 &= N(y_0),\\
A_1 &= y_1 N'(y_0)\\
A_2 &= y_2 N'(y_0) + \frac{1}{2!} y_1^2 N''(y_0)\\
A_3 &= y_3 N'(y_0) + y_1 y_2 N''(y_0) + \frac{1}{3!} y_1^3 N'''(y_0)\\
&\cdots\cdots\cdots\cdots\cdots\cdots\cdots\cdots\cdots\cdots
\end{aligned} \tag{7.52}$$

The convergence analysis of the solution and some recent modifications of ADM can be seen in (Almazmumy et al., 2012; Hosseini and Nasabzadeh, 2006).

Example 7.6

Solve the following IVP using ADM:

$$\frac{dy}{dx} + y^3 = 3, \quad y(0) = 0 \tag{7.53}$$

Solution: Take

$$L(y) = \frac{dy}{dx}; \; N(y) = y^3; \; f = 3 \tag{7.54}$$

Using Equation (7.44), we obtain

$$y(x) = y(0) + 3x - L^{-1}(y^3)$$

Let

$$y(x) = y_0 + \sum_{k=1}^{\infty} y_k(x)$$

be an approximate solution of the given problem

From Equation (7.50), $y_0 = 3x$. To find the subsequent terms, let us first construct the Adomian polynomials using Equation (7.53):

$$A_0 = N(y_0) = (3x)^3 = 27x^3$$

and

$$y_1 = -L^{-1}R(y_0) - L^{-1}(A_0) = -\int A_0\,dx = -\frac{27}{4}x^4$$

Similarly,

$$A_1 = y_1 N'(y_0) = 3y_0^2 y_1 = -\frac{729}{4}x^6$$

Hence,

$$y_2 = -L^{-1}R(y_1) - L^{-1}(A_1) = \frac{729}{28}x^7$$

$$y(x) = 3x - \frac{27}{4}x^4 + \frac{729}{28}x^7 - \frac{729}{560}x^9(54x+35)\ldots$$

is the required solution.

Example 7.7

Solve

$$y' = -y^2, \quad y(0) = 1$$

using ADM.

Solution: You may note that the exact solution to this IVP is

$$y(x) = \frac{1}{1+x}, \quad |x| < 1$$

Applying an ADM operator to the given equation, we obtain

$$y(x) = y(0) - L^{-1}(y^2) \tag{7.55}$$

Now, substituting

$$y = \sum_{k=0}^{\infty} y_k, \quad y^2 = \sum_{k=0}^{\infty} A_k \quad \& \quad y(0) = 1$$

in Equation (7.55), we have

$$\sum_{k=0}^{\infty} y_k = 1 - L^{-1}\left(\sum_{k=0}^{\infty} A_k\right)$$

Here,

$$y_0 = 1$$

and

$$y_{k+1} = -L^{-1}\left(\sum_{k=0}^{\infty} A_k\right), \quad k \geq 0$$

A simple and straightforward calculation of y_n's gives

$$y_1 = -x, \quad y_2 = x^2, \quad y_3 = x^3, \ldots$$

Hence, the required ADM approximate solution is

$$y(x) = 1 - x + x^2 - x^3 + \ldots$$

Figure 7.3 shows a comparison between an exact solution and the approximate solutions.

Solving Riccati's Equation Using the Adomian Decomposition Method

An equation of the form

$$\frac{dy}{dx} = P(x)y + Q(x)y^2 + T(x), \quad y(0) = S(x) \tag{7.56}$$

is called Riccati's equation, and $P(x)$, $Q(x)$, $T(x)$, and $S(x)$ are scalar functions.

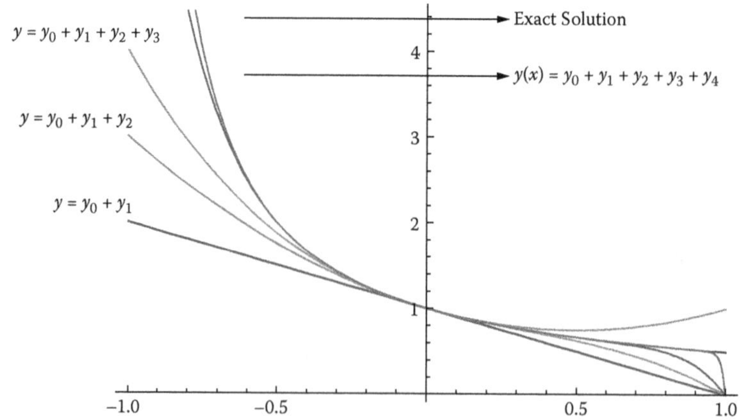

Figure 7.3 Comparison between exact solution and the approximate solutions.

Example 7.8

Solve

$$\frac{dy}{dx} = -\frac{x^2}{4}y^2 + x^3y + x^4 + 2, \quad y(0) = 0 \tag{7.57}$$

Solution: Rewrite Equation (7.57) as in Equation (7.56) to obtain

$$\frac{dy}{dx} = x^3y - \frac{x^2}{4}y^2 + x^4 + 2, \quad y(0) = 0 \tag{7.58}$$

Here,

$$P(x) = x^3, \quad Q(x) = -\frac{x^2}{4}, \quad T(x) = x^4 + 2, \quad S(x) = 0$$

Take

$$L(y) = \frac{dy}{dx}; \quad N(y) = \frac{x^2y^2}{4}; \quad R(y) = -x^3y \; f = x^4 + 2$$

Thus,

$$y(x) = \underbrace{y(0) + L^{-1}\left(x^4 + 2\right)}_{y_0} + L^{-1}\left(x^3y\right) - L^{-1}\left(\frac{x^2}{4}y^2\right) \tag{7.59}$$

Let

$$y = \sum_{k=0}^{\infty} y_k, \quad y^2 = \sum_{k=0}^{\infty} A_k \tag{7.60}$$

$$y_0 = y(0) + L^{-1}(x^4 + 2) = \int_0^x (x^4 + 2)dx = \frac{x^5}{5} + 2x \tag{7.61}$$

$$A_0 = N(y_0) = \frac{x^2}{4}\left(\frac{x^5}{5} + 2x\right)^2 \tag{7.62}$$

Then,

$$y_{k+1} = L^{-1}\left(x^3\sum_{k=0}^{\infty} y_k\right) - \frac{1}{4}L^{-1}\left(x^2\sum_{k=0}^{\infty} A_k\right), \quad k \geq 0 \tag{7.63}$$

Put $k=0$ in Equation (7.63) to obtain

$$y_1 = L^{-1}(x^3 y_0) - \frac{1}{4} L^{-1}(x^2 A_0) \tag{7.64}$$

$$L^{-1}\left(x^3 y_0\right) = \int_0^x \left(\frac{x^8}{5} + 2x^4\right) dx$$

$$L^{-1}\left(x^2 A_0\right) = \int_0^x x^2 \left(\frac{x^5}{5} + 2x\right)^2 dx$$

Now, substitute this in Equation (7.64) to obtain

$$y_1 = \frac{-x^{13}}{1300} + \frac{x^5}{5}.$$

Similarly, we obtain

$$y_2 = \frac{x^{21}}{156000} + \frac{3x^{17}}{4420} - \frac{x^{13}}{650}.$$

Now, the approximation $\varphi_n(x)$ for $y(x)$ for $n=2$ is

$$y(x) \approx \varphi_2(x) = y_0 + y_1 + y_2 + \ldots$$

$$\Rightarrow y(x) = 2x + \frac{2x^5}{5} - \frac{3x^{13}}{1300} + \frac{3x^{17}}{4420} + \frac{x^{21}}{156000} + \ldots$$

which is the required approximate solution to the given equation. This method gives us a better approximation, as can be seen Figure 7.4.

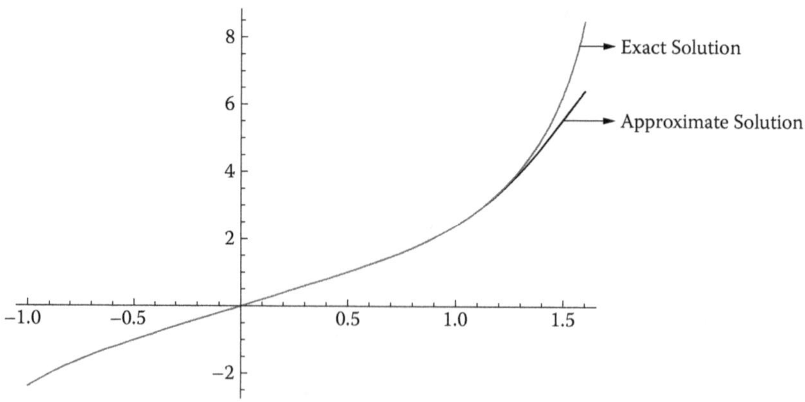

Figure 7.4 Comparison between exact solution and the truncated solution.

Note: From the work of Rao (2010), note that the absolute errors of the exact and Adomian approximate solutions are small for small values of x. Hence, the ADM has found approximate solutions of Riccati's differential equations.

Generalization of the Adomian Decomposition Method for Higher-Order Equations

For generalization of the ADM for higher-order equations (Evans and Raslan, 2004), consider the equation

$$L(y(x) - N(y(x)) = f, \quad 0 < x < 1, \quad y^{(i)}(0) = y_0^i, \quad i = 0,1,2,\ldots n-1 \tag{7.65}$$

where $L = \dfrac{d^n}{dx^n}$ is a differential operator, and N is a nonlinear operator.

Hence,

$$L^{-1}(.) = \int_0^x \int_0^{t_n} \int_0^{t_{n-1}} \cdots\cdots \int_0^{t_1} (.)\, dt_1 dt_2 \ldots dt_n$$

Operating with L^{-1} on Equation (7.65), it then follows that

$$L^{-1}L(y) - L^{-1}N(y) = L^{-1}f$$

As shown for a twofold method, a straightforward calculation yields

$$L^{-1}Ly = y(x) - \sum_{k=0}^{n-1} \frac{x^k}{k!} y^{(k)}(0)$$

(up to n terms), we obtain

$$y(x) - \sum_{k=0}^{n-1} \frac{x^k}{k!} y^{(k)}(0) = L^{-1}f + L^{-1}N(y)$$

$$\Rightarrow y(x) = \sum_{k=0}^{n-1} \frac{x^k}{k!} y^{(k)}(0) + L^{-1}f + L^{-1}N(y) \tag{7.66}$$

Let

$$\alpha_k = y^{(k)}(0)$$

and

$$y(x) = \sum_{n=0}^{\infty} y_n(x)$$

Then, Equation (7.66) takes the form

$$\sum_{n=0}^{\infty} y_n(x) = \sum_{k=0}^{n-1} \frac{x^k}{k!}\alpha_k + L^{-1}f + L^{-1}N(y)$$

where α_k's are constants that are given by the initial conditions.

The ADM assumes that the unknown function $y(x)$ can be expressed by an infinite series of the form

$$y_0(x) = \sum_{k=0}^{N-1} \frac{\alpha_k}{k!}x^k + L^{-1}f, \quad y_{n+1} = L^{-1}(A_n), \quad n = 0,1,2\ldots \qquad (7.67)$$

Using Equation (7.67), we can determine the components of $y_n(x), n \geq 0$ and hence the ADM solution.

Example 7.9

Solve

$$y^{(4)} + y\,y' - 3x^5 - 12 = 0, \quad y(0) = 0, \quad y''(0.5) = 3, \quad y'''(0.25) = 6,$$
$$y(1) = 1$$

Solution: Using Equation (7.66), these equations can be written as

$$y(x) = y(0) + x\alpha_1 + \frac{x^2}{2!}\alpha_2 + \frac{x^3}{3!}\alpha_3 + L^{-1}(3x^5 + 12) - L^{-1}(y\,y') \qquad (7.68)$$

Here,

$$L^{-1}(3x^5+12)=\int_0^x\int_0^x\int_0^x\int_0^x(3x^5+12)dx_1dx_2dx_3dx=\frac{x^4}{2}+\frac{x^9}{1008} \quad (7.69)$$

Now, on substituting $y(0)=0$ and Equation (7.69) in Equation (7.68), we obtain

$$y(x)=\underbrace{0+x\alpha_1+\frac{x^2}{2}\alpha_2+\frac{x^3}{6}\alpha_3+\frac{x^4}{2}+\frac{x^9}{1008}}_{y_0}-L^{-1}(y\,y') \quad (7.70)$$

where

$$y_0=x\alpha_1+\frac{x^2}{2}\alpha_2+\frac{x^3}{6}\alpha_3+\frac{x^4}{2}+\frac{x^9}{1008} \quad (7.71)$$

$$y_{n+1}=-L^{-1}(y_n.y_n'),\quad n\geq 0 \quad (7.72)$$

If we put $n=0$, we obtain

$$y_1=\frac{\alpha_1^2x^5}{120}+\frac{\alpha_1\alpha_2x^6}{240}+\frac{\alpha_2x^7}{1680}+\frac{\alpha_1\alpha_3x^7}{1260}+\frac{\alpha_1x^8}{672}+\frac{\alpha_2\alpha_3x^8}{4032}$$
$$+\frac{\alpha_2x^9}{2016}+\frac{\alpha_3^2x^9}{36288}+\frac{\alpha_3x^{10}}{8640}+\frac{x^{11}}{7920}+\frac{\alpha_1x^{13}}{1729728}+\frac{\alpha_2x^{14}}{4402944}$$
$$+\frac{\alpha_3x^{15}}{16511040}+\frac{x^{16}}{6773760}+\frac{x^{21}}{16216381440} \quad (7.73)$$

The required solution obtained by substituting y_0 and y_1 in $y(x)$ is

$$y(x)=x\alpha_1+\frac{x^2}{2}\alpha_2+\frac{x^3}{6}\alpha_3+\frac{x^4}{2}+\frac{x^9}{1008}$$
$$-\left(\begin{array}{l}\dfrac{\alpha_1^2x^5}{120}+\dfrac{\alpha_1\alpha_2x^6}{240}+\dfrac{\alpha_2x^7}{1680}+\dfrac{\alpha_1\alpha_3x^7}{1260}+\dfrac{\alpha_1x^8}{672}+\dfrac{\alpha_2\alpha_3x^8}{4032}\\ +\dfrac{\alpha_2x^9}{2016}+\dfrac{\alpha_3^2x^9}{36288}+\dfrac{\alpha_3x^{10}}{8640}+\dfrac{x^{11}}{7920}+\dfrac{\alpha_1x^{13}}{1729728}+\dfrac{\alpha_2x^{14}}{4402944}\\ +\dfrac{\alpha_3x^{15}}{16511040}+\dfrac{x^{16}}{6773760}+\dfrac{x^{21}}{16216381440}\end{array}\right)-\ldots.$$

(7.74)

Applying

$$y''(0.5) = 3, \quad y'''(0.25) = 6 \quad \& \quad y(1) = 1$$

to Equation (7.74), we obtain $\alpha_1, \alpha_2, \alpha_3$, where

$$\alpha_1 = 0.499276782, \quad \alpha_2 = -0.000497001, \quad \alpha_3 = -0.00012207$$

By substituting the values of $\alpha_1, \alpha_2, \alpha_3$ in Equation (7.74), we obtain the required approximate solution to the given problem.

Modified Adomian Decomposition Method for Singular IVP Problems for ODEs

The discussion now turns to the modified ADM for singular IVP problems for ODEs (Hasan and Zhu, 2008). Let us consider the singular IVP in the second-order ODE of the form

$$y'' + \frac{2}{x}y' + f(x, y) = g(x), \quad y(0) = A, \quad y'(0) = B \tag{7.75}$$

where $f(x,y)$ is a real valued nonlinear function, $g(x)$ is the given function, and A and B are given initial values.

Here, we propose the new differential operator as follows:

$$L = x^{-1}\frac{d^2}{dx^2}xy \tag{7.76}$$

where

$$L = \frac{d^2 y}{dx^2}$$

Equation (7.75) can be written as

$$Ly = g(x) - f(x,y) \tag{7.77}$$

Placing an operator L^{-1} on both sides of Equation (7.77), we obtain

$$L^{-1}L(y(x)) = L^{-1}(g(x)) - L^{-1}(f(x, y)) \tag{7.78}$$

From Equation (7.75),

$$L(y(x)) = \left(y'' + \frac{2}{x} y' \right)$$

Here, L^{-1} is considered a twofold integral operator as follows:

$$L^{-1}(.) = x^{-1} \int_0^x \int_0^x x(.)\, dx\, dx \tag{7.79}$$

From Equation (7.75), we obtain

$$L^{-1}(.) = L^{-1}\left(y'' + \frac{2}{x} y' \right)$$

then, Equation (7.79) becomes

$$\begin{aligned} L^{-1}\left(y'' + \frac{2}{x} y' \right) &= x^{-1} \int_0^x \int_0^x x\left(y'' + \frac{2}{x} y' \right) dx\, dx \\ &= x^{-1} \int_0^x \left(xy' + y - y(0) \right) dx = y(x) - y(0) \end{aligned} \tag{7.80}$$

Substituting Equation (7.80) in Equation (7.78), we obtain

$$y(x) - y(0) = L^{-1}(g(x)) - L^{-1}(f(x, y))$$

Hence,

$$y(x) = y(0) + L^{-1}(g(x)) - L^{-1}(f(x, y)) \tag{7.81}$$

By introducing Adomian decomposition polynomials to Equation (7.81), we obtain the solution $y(x)$ of this equation as in Equation (7.66) where

$$y(x) = \sum_{n=0}^{\infty} y_n(x) \quad \text{and} \quad f(x, y) = \sum_{n=0}^{\infty} A_n \tag{7.82}$$

Substituting Equation (7.82) and the initial condition $y(0) = A$ of Equation (7.75) Equation (7.76) can be written as

$$\sum_{n=0}^{\infty} y_n = A + L^{-1}(g(x)) - L^{-1}\left(\sum_{n=0}^{\infty} A_n\right) \tag{7.83}$$

The required Adomian polynomials $A_0, A_1, A_2 \ldots$ can be obtained as in Equation (7.52) and $y_n(x)$ determined as

$$y_0 = A + L^{-1}(g(x)) \quad \text{and} \quad y_{k+1} = -L^{-1}(A_k), \quad k \geq 0 \tag{7.84}$$

Hence, the nth term approximately can be obtained as

$$\phi_n(x) = \sum_{n=0}^{n-1} y_k$$

Example 7.10

Solve the nonlinear singular IVP

$$y'' + \frac{2}{x}y' + y^3 = 6 + x^4, \quad y(0) = 0, \quad y'(0) = 0$$

Solution: This equation can be written as

$$L(y) = 6 + x^4 - y^3$$
$$\Rightarrow \; y(x) = y(0) + L^{-1}(6 + x^4) - L^{-1}(y^3)$$

Using Equation (7.79), the value of $L^{-1}(6 + x^4)$ can be obtained. Hence,

$$L^{-1}(6 + x^4) = x^{-1}\int_0^x\int_0^x x(6 + x^4)\,dx\,dx$$
$$= x^2 + \frac{x^6}{42}$$

Thus,

$$y(x) = x^2 + \frac{x^6}{42} - L^{-1}(y^3)$$

By ADM (Wawaz, 1999), we can split $x^2 + \frac{x^6}{42}$ into two parts:

$$y_0 = x^2$$

and

$$y_{k+1} = \frac{x^6}{42} - L^{-1}(A_k)$$

We use the Adomian polynomial for the nonlinear term to obtain $y_1, y_2, \ldots$.

Now, putting $k = 1$ in y_{k+1}, we obtain

$$y_1 = \frac{x^6}{42} - \frac{1}{x}\int_0^x\int_0^x x(x^2)^3\,dx\,dx = 0$$

$$y_{k+1} = 0,\ k \geq 0$$

where

$$y_1 = y_2 = y_3 = \ldots = 0$$

Thus, the exact solution is $y(x) = x^2$.

Exercise Problems

1. $$\frac{dy}{dx} = -y^2 + 1, \quad y(0) = 0$$

2. $$\frac{dy}{dx} = x^3y^2 - 2x^4y + x^5 + 1, \quad y(0) = 0$$

3. $$\frac{d^3y}{dx^3} = -y(x) - y(x - 0.3) + e^{-x+0.3}, \quad 0 \leq x \leq 1, \quad y(0) = 1,$$
$$y'(0) = -1, \quad y''(0) = 1, \quad y(x) = e^{-x}, x \leq 0$$

4. $$y^{(4)} + yy'' - 4x^7 - 24 = 0, \quad y(0) = 0, \quad y''(0.5) = 3,$$
$$y'''(0.25) = 6, \quad y(1) = 1$$

5. $$y' = e^y + 1, \quad y(0) = 1$$

6. Solve the nonlinear singular IVP by modified ADM:

$$y'' + \frac{2}{x}y' + y^3 = 6 + x^6, \quad y(0) = 0, \quad y'(0) = 0$$

7. Solve this equation using modified ADM:

$$y'' + \frac{2}{x}y' + y = 6 + 12x + x^2 + x^3, \quad y(0) = 0 = y'(0)$$

Hints:

1. $y_0 = x, \quad y_1 = \frac{-x^3}{3}, \quad y_2 = \frac{2x^5}{15}, \quad y_3 = \frac{-17x^7}{315}, \quad y_4 = \frac{62x^9}{2835}, \ldots$

2. $y_0 = \frac{x^6}{6} + x, \quad y_1 = \frac{x^{16}}{576} - \frac{x^6}{6}, \quad y_2 = \frac{x^{26}}{44928} - \frac{x^{16}}{288}, \ldots$

3. $y_{n+1} = \int_0^x \int_0^x \int_0^x \left(-y_n(x) - y_n(x - 0.3)\right) dx\, dx\, dx$

   and

$$y_0 = \alpha_0 + x\alpha_1 + \frac{x^2}{2}\alpha_2 + e^{-x+0.3}$$

   where

$$\alpha_0 = 1 + e^{0.3}, \quad \alpha_1 = -1 - e^{0.3}, \quad \alpha_2 = 1 + e^{0.3}$$

4. $y_{n+1} = -L^{-1}(y_n . y_n'), \quad n \geq 0,$

$$y = \alpha_0 + \alpha_1 x + \alpha_2 \frac{x^2}{2} + \alpha_3 \frac{x^3}{6} + \frac{1}{1980}x^{11} + x^4$$

   where

$$\alpha_0 = 0, \quad \alpha_1 = -0.1674e-10, \quad \alpha_2 = -0.3665e-13,$$
$$\alpha_3 = -0.1635e-13$$

5. $y_0 = (1 + x), \quad y_1 = e^{1+x}, \quad y_2 = \frac{e^{2(1+x)}}{2}, \quad y_3 = \frac{e^{3(1+x)}}{3}, \ldots$

6. $y_0 = x^2, \quad y_{k+1} = 0$

7. $y_0 = x^2 + x^3, \quad y_{k+1} = 0$

Applications

G.A. Afrouzi and S. Khademloo. On Adomian decomposition method for solving reaction diffusion equation. *International Journal of Nonlinear Science,* Vol. 2, No. 1, pp. 11–15, 2006.

P.K. Bera and J. Datta. Linear delta expansion technique for the solution of anharmonic oscillations. *Pramana—Journal of Physics,* Vol. 68, pp. 117–122, 2007.

M.P. Blencowe and A.P. Korte. Applying the linear δ-expansion to disordered systems. *Physics Review,* B56, 9422, 1997.

A. Cheniguel and A. Ayadi. Solving heat equation by the Adomian decomposition method. *Proceedings of the World Congress on Engineering 2011,* Vol. 1, WCE 2011, July 6–8, 2011, London.

M. Danesh and M. Safari. Application of Adomian's decomposition method for the analytical solution of space fractional diffusion equation. *APM,* Vol. 1, pp. 345–350, 2011.

J.-S. Duan, R. Rach, D. Baleanu, and A.-M. Wazwaz. A review of the Adomian decomposition method and its applications to fractional differential equations. *Communications in Fractional Calculus,* Vol. 3, pp. 73–99, 2012.

S.A. El-Wakil, M.A. Abdou, and A. Elhanbaly. Adomian decomposition method for solving the diffusion–convection–reaction equations. *Applied Mathematics and Computation,* Vol. 177, pp. 729–736, 2006.

J.M. Machado, S.L.L. Verardi, and Y. Shiyou. An application of the Adomian's decomposition method to the analysis of MHD duct flows. *IEEE Transactions on Magnetics,* Vol. 41, pp. 1588–1591, May 2005.

M. Tatari, M. Dehghan, and M. Razzaghi. Application of the Adomian decomposition method for the Fokker–Planck equation. *Mathematical and Computer Modelling,* Vol. 45, pp. 639–650, 2007.

R.A. VanGorder. δ-Expansion method for nonlinear stochastic differential equations describing wave propagation in a random medium. *Physical Review. E, Statistical, Nonlinear, and Soft Matter Physics,* Vol. 82, 056712, 2010.

L.I. Xiu-mei, W.U. Feng, S.U. Jin-ling. Lyapunov method for the analysis of structural responses. *Engineering Mechanics,* Issue 11, pp. 16–21, 2010.

Bibliography

Adomian decomposition method. 2014. http://en.wikipedia.org/wiki/Adomian_decomposition_method.

G. Adomian. Nonlinear stochastic differential equations. *Journal of Mathematical Analysis and Applications,* Vol. 55, pp. 441–452, 1976.

G. Adomian. A review of the decomposition method in applied mathematics. *Journal of Mathematical Analysis and Applications*, Vol. 135, pp. 501–544, 1988.

G. Adomian. A review of the decomposition method and some recent results for nonlinear equation. *Mathematical and Computer Modeling*, Vol. 13, pp. 17–43, 1990.

G. Adomian. *Solving Frontier Problems of Physics: The Decomposition Method*. Dordrecht, the Netherlands: Kluwer Academic, 1994.

G. Adomian and G.E. Adomian. A global method for solution of complex systems. *Mathematical Modeling*, Vol. 5, pp. 521–568, 1984.

M. Almazmumy, F.A. Hendi, H.O. Bakodah, and H. Alzumi. Recent modifications of Adomian decomposition method for initial value problems in ordinary differential equations. *American Journal of Computational Mathematics*, Vol. 2, pp. 228–234, 2012.

Basic ideas and brief history of the homotopy analysis method. n.d. http://numericaltank.sjtu.edu.cn/BasicIdea-BriefHistory/BasicIdea-BriefHistory.pdf.

Y. Cherrault. Convergence of Adomian decomposition method. *Applied Mathematics and Computation*, pp. 101–106, 2006.

D.J. Evans and K.R. Raslan. The Adomian decomposition method for solving delay differential equation. *International Journal of Computer Mathematics*, Vol. 82, pp. 49–54, 2005.

Y. Q. Hasan and L. M. Zhu. Modified Adomian decomposition method for singular initial value problems in the second order ordinary differential equations. *Surveys in Mathematics and Its Applications*, Vol. 3, pp. 183–193, 2008.

M.M. Hosseini and H. Nasabzadeh. On the convergence of Adomian decomposition method. *Applied Mathematics and Computation*, Vol. 182, pp. 536–543, 2006.

S. Liang and D.J. Jeffrey. Comparison of homotopy analysis method and homotopy perturbation method through an evolution equation. *Communications in Nonlinear Science and Numerical Simulation*, Vol. 16, pp. 4057–4064, 2009.

T.R. Ramesh Rao. The use of Adomian decomposition method for solving generalized Riccati differential equations. *Proceedings of 6th IMT-GT Conference on Mathematics, Statistics and Its Applications* (ICMSA 2010), Universiti Tunku Abdul Rahman, Kuala Lumpur, Malaysia, pp. 935–941, 2010.

A.M. Wawaz. A reliable modification of Adomian decomposition method. *Applied Mathematics and Computation*, Vol. 102, 77–86, 1999.

8

Homotopy Methods

Introduction

In this chapter, we present homotopy methods that are of fairly recent origin. These methods also provide us approximate analytical solutions that are in terms of series of functions of the independent variable. The functions in the solution constitute a set of base functions (which are linearly independent functions). These methods are applicable to problems modeled by linear as well as weakly/strongly nonlinear ordinary differential equations. The methods considered here are

1. The homotopy analysis method (HAM) initiated by Liao
2. The homotopy perturbation method (HPM) proposed by He
3. The optimal homotopy asymptotic method (OHAM) discussed by Marinca and Herisanu

These methods and a number of offshoots have been extensively used recently to solve a wide variety of problems (see the Applications section at the end of the chapter).

Homotopy Analysis Method

Liao developed HAM as early as 1992. Based on the concept of homotopy in topology, HAM overcomes the limitations of the methods discussed in previous chapters as follows:

1. The perturbation method is useful when the given equation has a small or large perturbation parameter; this may not be the case with a wide variety of problems, and thus the method is not applicable.
2. Although the Lyapunov artificial parameter method and the δ expansion method are applicable for problems not involving

the small parameter, it is necessary to introduce an artificial small parameter into the given equation and find its solution in terms of a series in this artificial parameter. The major limitation of these methods is that there is no theory that guides us regarding where to put the artificial parameter in a better place for better approximation.

3. Although the Adomian decomposition method is useful for a number of problems, it does not provide the methodology to choose suitable base functions, and there is no way for controlling the rate of convergence of the solution.

Liao's HAM provides for a choice of different base functions other than polynomials, which of course are chosen according to the given problem. It provides a control on the convergence region and the rate of convergence of the solution, which is achieved by means of a convergence control parameter and an auxiliary function that are introduced to the problem formulation.

Liao proved that "the HAM logically contains Lyapunov's artificial small parameter method, the δ-expansion method and Adomian decomposition method. It unifies these nonperturbation methods and is more general than them" (Liao, 2003, chapter 4). He also showed that this method always provides analytical approximations efficiently that are accurate for all physical parameters.

Method

Consider a nonlinear differential equation of the form

$$N(u(x)) = 0 \tag{8.1}$$

where N is a nonlinear operator, x is the independent variable, and $u(x)$ is the unknown function. Let $u_0(x)$ be the initial approximation of the exact solution $u(x)$ and L be an auxiliary linear operator with the property that

$$L(f) = 0 \text{ when } f = 0. \tag{8.2}$$

In this method, we construct the homotopy, which is a continuous mapping $H : u(x) \rightarrow \phi(x;q)$, defined as

$$H(\phi(x;q);q) = (1-q)L(\phi(x;q) - u_0(x)) - \hbar H(x)qN(\phi(x;q)) \tag{8.3}$$

Here, $H(x)$ is an auxiliary function; h is an auxiliary parameter called the convergence control parameter; $q \in [0,1]$ is an embedding parameter; and $\phi(x;q)$ is the approximate solution to the given problem. We notice from Equation (8.3) that the solution obtained using this method depends on four important factors: the initial approximation $u_0(x)$, the linear operator L, the auxiliary function $H(x)$, and the auxiliary parameter h.

When $q = 0$ and when the homotopy defined by Equation (8.3) is taken to be zero, we obtain the zeroth-order deformation equation given by

$$L(\phi(x;0) - u_0(x)) = 0. \tag{8.4}$$

In view of the linearity of the operator L, the zeroth deformation equation is given by

$$\phi(x;0) = u_0(x). \tag{8.5}$$

Now, when $q = 1$, Equation (8.3) takes the form

$$N(\phi(x;1)) = 0.$$

This equation is the same as the given equation provided

$$\phi(x;1) = u(x). \tag{8.6}$$

This shows that, as the embedded parameter q varies from 0 to 1, $\phi(x;q)$ varies from the initial guess $u_0(x)$ [as is seen in Equation (8.5)] to the exact solution $u(x)$ [as seen Equation (8.6)].

Let us now define the mth-order deformation derivatives as

$$u_0^{[m]}(x) = \left.\frac{\partial^m}{\partial q^m}(\phi(x;q))\right|_{q=0}. \tag{8.7}$$

Then, using Taylor's theorem, $\phi(x;q)$ can be expanded as a power series of q as

$$\phi(x;q) = \phi(x;0) + \sum_{m=1}^{\infty} \frac{u_0^{[m]}(x)}{m!} q^m. \tag{8.8}$$

Writing $u_m(x) = \frac{u_0^{[m]}(x)}{m!}$ and using Equation (8.5), Equation (8.8) takes the form

$$\phi(x;q) = u_0(x) + \sum_{m=1}^{\infty} u_m(x)q^m. \tag{8.9}$$

With a suitable choice for the initial guess, the auxiliary linear operator, the convergence control parameter, and the auxiliary function, Liao proved that the power series solution converges for $q = 1$ (see Liao, 2003).

Now, to find the solution using Equation (8.9), we need to find the functions $u_m(x)$ for $m = 1, 2, 3, \ldots$.

Liao (2003) derived that these functions are given by the mth-order deformation equation, defined as follows:

$$L(u_m(x) - \chi_m u_{m-1}(x)) = \hbar H(x) R_m(\vec{u}_{m-1}(x)) \tag{8.10}$$

where

$$R_m(\vec{u}_{m-1}(x)) = \frac{1}{(m-1)!}\left(\frac{\partial^{m-1}}{\partial q^{m-1}}(N(\phi(x;q)))\right)_{q=0} \tag{8.11}$$

and

$$\chi_m = \begin{cases} 0, & m \leq 1 \\ 1, & otherwise \end{cases}. \tag{8.12}$$

After determining $u_m(x)$ for $m = 1, 2, \ldots$, an approximate solution to the problem given in Equation (8.1) is

$$u(x) \approx u_0(x) + \sum_{m=1}^{\infty} u_m(x). \tag{8.13}$$

For details on the convergence aspect of the series solution, consult the related topics from the work of Liao (2003).

Rules for the Selection of the Different Functions and the Parameters Described in the Definition Given by Equation (8.3)

The great flexibility in HAM is that the solution to the given nonlinear problem can be expressed in terms of different sets of base functions $\{e_n(x)/n = 1,2,3...\}$. To choose the appropriate set, we use a rule called the "rule of solution expression." This rule states that the base functions have to be determined from the physical characteristics of the problem or the boundary or initial conditions specified. Once the set of base functions is identified, we assume the solution to the given problem is

$$u(x) = \sum_{m=0}^{\infty} c_m e_m(x)$$

where the c_m's are unknown coefficients to be determined.

Further, the initial approximation is to be taken as

$$u_0(x) = \sum_{m=0}^{M} a_m e_m(x)$$

where M is an integer and the a_m's are unknown coefficients to be determined.

For instance, consider the problem

$$u'' + \varepsilon(u - u^3) = 0, u(0) = u(\pi) = 0.$$

The initial conditions guide us to take the base functions as $\{\sin(2m+1)x/m = 0,1,2\ldots\}$.

Thus,

$$u(x) = \sum_{m=0}^{\infty} c_m \sin(2m+1)x$$

and

$$u_0(x) = \sum_{m=0}^{M} a_m \sin(2m+1)x.$$

Also, the linear operator must be chosen in such a way that the solution of the equation

$$L(w(x)) = 0$$

must be expressed by a sum of the base functions as

$$w(x) = \sum_{n=0}^{M_0} b_n e_n(x)$$

where M_0 is an integer determined by the highest-order derivative in the linear operator and the b_n's are unknown coefficients to be determined.

Next comes the "rule of ergodicity," which states that as the order of approximation tends to infinity, each base function should appear in the solution expression and each coefficient should be able to be modified. Using this and the previous rules, the auxiliary function $H(x)$ can be uniquely determined.

We see from Equation (8.10) that the higher-order deformation equations transform the given nonlinear problem into an infinite number of linear subproblems. The "rule of solution existence" imposes further restrictions on the choice of the initial approximation, auxiliary linear operator, and the auxiliary function. This rule states that the auxiliary function and the linear operator are to be so chosen that all the higher-order deformation equations must have solutions, provided the original problem has a solution.

Examples 8.1

Solve the equation

$$(1+\varepsilon u)\frac{du}{dx} + u = 0, \quad u(0) = 1$$

using HAM. (This equation is used in the study of the cooling of a lumped system with variable specific heat.)

Solution: In view of the given condition and the physical description of the problem, the suitable form of base functions is $\{e^{-nx}, n = 1,2,3\ldots\}$.

The solution $u(x)$ now takes the form

$$u(x) = \sum_{n=1}^{\infty} a_n e^{-nx}$$

Choose $L(u) = u' + u$

$$N(u) = (1+\varepsilon u)u' + u$$

Construct the homotopy:

$$H(\phi(x;q);q) = (1-q)L(\phi(x;q) - u_0(x)) - \hbar H(x)qN(\phi(x;q))$$

Let

$$u_0(x) = e^{-x}$$

Hence, $\phi(x;0) = e^{-x}$ [from Equation (8.5)].

The first-order deformation equation is

$$L(u_1(x) - \chi_1 u_0(x)) = \hbar H(x) R_1(\vec{u}_0(x))$$

$$\Rightarrow L(u_1(x)) = \hbar \frac{1}{0!}(N(\phi(x,q)))_{q=0}$$

and

$$\phi(x;q) = u_0(x)$$

Hence, after straightforward calculations as described in the method, we obtain

$$u_1' + u_1 = -\varepsilon \hbar H(x) e^{-2x}; u_1(0) = 0$$

Using the rule of ergodicity and the rule of solutions, we take $H(x) = 1$.

Solving the previous initial value problem (IVP), we obtain

$$u_1(x) = \varepsilon \hbar (e^{-2x} - e^{-x})$$

Similarly, we obtain that

$$u_2(x) = e^{-x}\left(-\varepsilon\hbar(1+\hbar) + \frac{\varepsilon^2}{2}\hbar^2\right) + e^{-2x}(\varepsilon\hbar(1+\hbar) - 2\varepsilon^2\hbar^2) + e^{-3x}\left(\frac{3}{2}\varepsilon^2\hbar^2\right)$$

Thus, the solution is

$$u(x) = u_0(x) + u_1(x) + u_2(x) + \cdots$$

Note: As can be seen in the example, the solution involves the parameter $\hbar$. It was proved by Liao that this parameter often affects the convergence region and the rate of convergence of the solution series; hence, $\hbar$ is to be determined in such a way that the series solution converges faster and has a larger region of convergence (Liao, 2003). To achieve this, Liao devised a certain class of curves called $\hbar$ curves.

$\hbar$ **curves:** We know that most of the problems contain some important physical parameters, such as the wall skin friction, the frequency

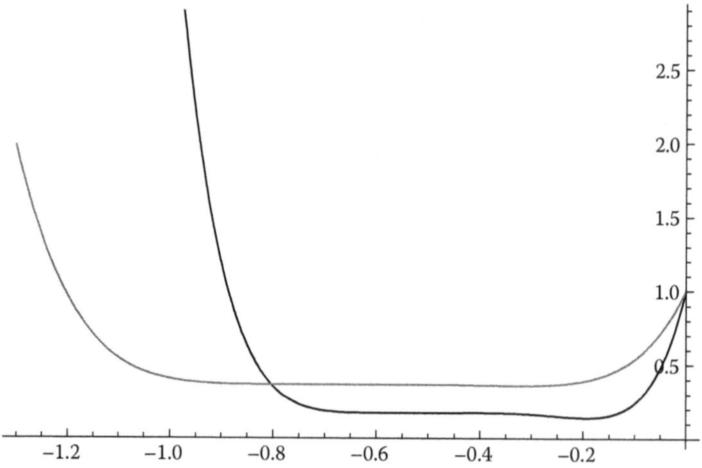

Figure 8.1 Plot of $u''(0)$ versus $\hbar$ for $\varepsilon = 0.4$ (gray curve), 0.8 (black curve).

of a nonlinear oscillator, and so on. Now that the homotopy solutions involve the convergence control parameter $\hbar$, so do these physical parameters. The curves obtained by plotting such parameters (evaluated at $x = 0$) as a function of $\hbar$ are called $\hbar$ curves. Liao has proved that if the solution to the given problem is unique, then all these curves converge to the same value; hence, there exists a horizontal line segment that gives the range of $\hbar$ for which the series solution is convergent. Now, for any value of $\hbar$ in this range, called the valid range, the corresponding solution series converges.

Let us find the valid range of $\hbar$ for the previous example: To find the valid range for $\hbar$, let us plot the $\hbar$ curves of $u''(0)$. From Figure 8.1, we can see that the valid region is roughly $-0.7 < \hbar < -0.3$.

Example 8.2

Solve the equation

$$\frac{du}{dx} - 2xu = 0, \quad u(0) = 1$$

using HAM.

Solution: Choose $L(u) = u'$

$$N(u) = u' - 2xu$$

$$H(x) = 1$$

Construct the homotopy:

$$H(\phi(x;q);q) = (1-q)L(\phi(x;q) - u_0(x)) - hqN(\phi(x;q))$$

Choose $u_0(x) = 1$. Then,

$$\phi(x;0) = 1$$

The first-order deformation equation is

$$L(u_1(x) - \chi_1 u_0(x)) = hR_1(\vec{u}_0(x))$$

$$\Rightarrow L(u_1(x)) = h\frac{1}{0!}(N(\phi(x,q)))_{q=0}$$

and

$$\phi(x;q) = u_0(x)$$

Hence,

$$u_1' = -2xh; \quad u_1(0) = 0$$

solving this, we obtain

$$u_1(x) = -x^2 h$$

Similarly, we obtain

$$u_2(x) = \frac{1}{2}\ h\ x^2\ (-2 - 2\ h + h\ x^2)$$

Thus, the solution is

$$u(x) = u_0(x) + u_1(x) + u_2(x) + \cdots$$

To find the valid range for h, we shall plot the h curves of $u''(0)$. From Figure 8.2, we can see that the valid range is roughly $-1.1 < h < -0.8$.

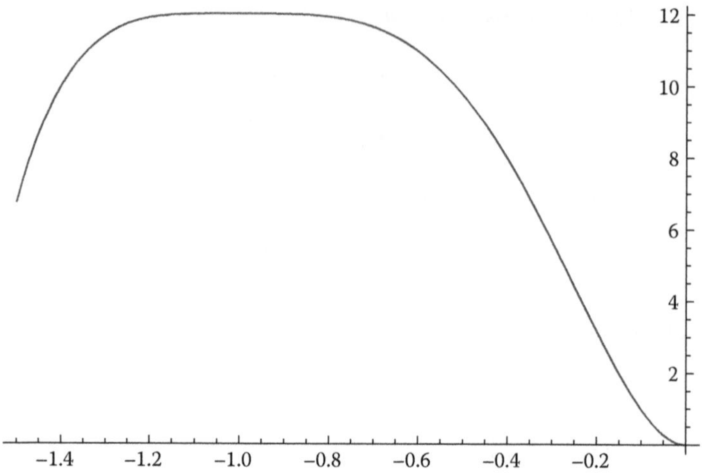

Figure 8.2 Plot of $u''(0)$ versus h.

Example 8.3

Solve the equation

$$u'' - \varepsilon u^n = 0, \quad u(0) = 0.25, \quad u'(0) = 0$$

using HAM.

Solution: Choose $L(u) = u''$:

$$N(u) = u'' - \varepsilon u^n$$

$$H(x) = 1$$

The base functions are

$$\{x^m / m = 1, 2, \ldots\}$$

Construct the homotopy:

$$H(\phi(x;q);q) = (1-q)L(\phi(x;q) - u_0(x)) - hqN(\phi(x;q))$$

Let $u_0(x) = 0.25$. Hence,

$$\phi(x;0) = 0.25$$

The first-order approximation:

$$L(u_1(x) - \chi_1 u_0(x)) = hR_1(\vec{u}_0(x))$$

$$\Rightarrow L(u_1(x)) = h\frac{1}{0!}(N(\phi(x,q)))_{q=0}$$

and

$$\phi(x;q) = u_0(x)$$

Hence,

$$u_1'' = -\hbar\varepsilon(0.25)^n; \quad u_1(0) = 0, \quad u_1'(0) = 0$$

Solving this, we have

$$u_1(x) = -\frac{\hbar\varepsilon(0.25)^n x^2}{2}$$

Similarly, we obtain

$$u_2(x) = \frac{1}{6}\hbar\varepsilon(0.25)^n\left(-3(1+\hbar)x^2 + \frac{1}{4}(0.25)^{n-1}\hbar\varepsilon n x^4\right)$$

Thus, the solution is

$$u(x) = u_0(x) + u_1(x) + u_2(x) + \cdots$$

To find the valid range for $\hbar$, we have to plot the $\hbar$ curves of $u''(0)$. From Figure 8.3, we can see that the valid range is roughly $-1.1 < \hbar < -0.5$. The graphs are drawn by evaluating $u(x)$ to a third approximation. A better result can be obtained by finding higher-order approximations to the solution.

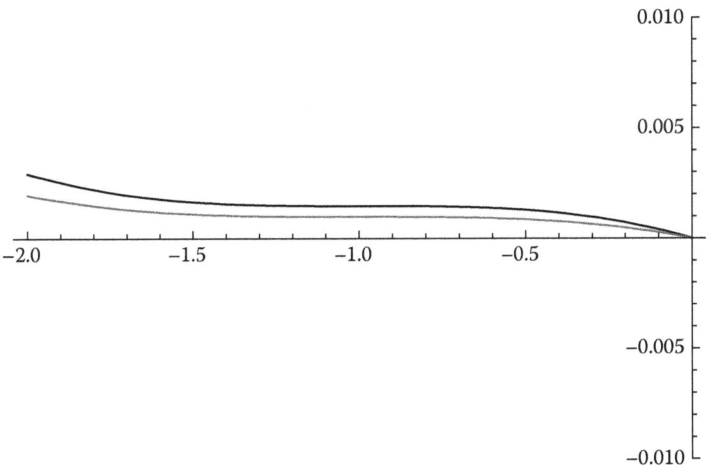

Figure 8.3 Plot of $u''(0)$ versus h for $\varepsilon = 1.0$ (gray curve), 1.5 (black curve).

Homotopy Perturbation Method

Consider the nonlinear differential equation given by

$$L(u(x)) + N(u(x)) + g(x) = 0, \quad B(u(x)) = 0 \tag{8.14}$$

where L is a linear operator, x is an independent variable, $u(x)$ is an unknown function, $g(x)$ is a known function, N is a nonlinear operator, and $B(u)$ is a boundary operator or an operator related to initial conditions.

By the method of HPM, we first construct a family of equations:

$$H(u(x), p) = (1 - p)\{L(u(x)) + L(u_0(x))\}$$
$$- p\{L(u(x)) + g(x) + N(u(x))\} = 0$$
$$B(u(x)) = 0 \tag{8.15}$$

where p lies in [0,1] and is an embedding parameter. Here, $u_0(x)$ is an initial approximation. When $p = 0$, $p = 1$, it holds that

$$H(u(x), 0) = L(u(x)) + L(u_0(x)) = 0$$
$$H(u(x), 1) = L(u(x)) + g(x) + N(u(x)) = 0 \tag{8.16}$$

Expanding $u(x, p)$ in a series with respect to p, we have

$$u(x, p) = u_0(x) + \sum_{k=1}^{\infty} u_k(x) p^k \tag{8.17}$$

Substituting Equation (8.17) in Equation (8.15) and comparing the coefficients of different powers of p, we obtain a set of differential equations with initial or boundary conditions. Solving these differential equations, we obtain $u_1(x), u_2(x), u_3(x) \ldots$, and the mth approximate solution of Equation (8.15) can be written as follows:

$$\tilde{u}(x) = u_0(x) + \sum_{k=1}^{m} u_k(x) \tag{8.18}$$

Note: Liao (2005) has shown that this HPM is a special case of HAM with $\hbar = -1$ and $H(x) = 1$ and pointed out that the convergence of the solution depends on only two factors: the initial guess $u_0(x)$ and the auxiliary linear operator L, unlike the HAM, for which we have four factors that influence the solution. Hence, the two factors are good enough for faster convergence of the solution. Liao also claimed that even if the initial guess for HAM is bad, the method still provides faster-converging solutions because of the convergence control parameter h. But, this is not true for the HPM solutions. Further, he said that for some strongly nonlinear problems (those that have multiple solutions, discontinuations, etc.), it is not possible to find good initial guesses, which is a major drawback for using HPM (Liao, 2005). These complications in the problem together with a bad guess of the initial solution may sometimes lead to divergent solutions, as shown by Liang and Jeffrey (2009). However, the major advantage of the method is that we can work easily for the problems that are weakly nonlinear, as seen in the next examples.

Example 8.4

Solve

$$y' - 2xy = 0, \quad y(0) = 1$$

using HPM.

Solution: Choose

$$L(y) = y'$$

and

$$N(y) = -2xy.$$

In view of the given initial condition,

$$y_0 = 1.$$

Construct the homotopy as [using Equation (8.15)]

$$H(y, p) = (1 - p)(y' - 1) + p(y' - 2xy)$$

Let

$$y(x, p) = y_0 + y_1 p + y_2 p^2 + \cdots$$

Substituting y in this homotopy and comparing the coefficients of like powers of p, we have

Coefficient of p:

$$y_1' = 2xy_0, \quad y_1(0) = 0$$

Coefficient of p^2:

$$y_2' = 2xy_1, \quad y_2(0) = 0$$

and so on.

Solving these IVPs, we have

$$y_1(x) = x^2$$

$$y_2(x) = \frac{x^4}{2}$$

and so on.

Hence,

$$y(x) = 1 + x^2 + \frac{x^4}{2!} + \cdots$$

The exact solution to this problem is $y(x) = e^{x^2}$.

Example 8.5

Solve

$$y' = 1 + y^2, \quad y(0) = 0.$$

Solution: Choose

$$L(y) = y'$$

and

$$N(y) = -1 - y^2.$$

In view of the given initial condition,

$$y_0 = 0.$$

Construct the homotopy as [using Equation (8.15)]

$$H(y, p) = (1 - p)(y') + p(y' - y^2 - 1)$$

Let

$$y(x,p) = y_0 + y_1 p + y_2 p^2 + \cdots$$

Substituting y in this homotopy and comparing the coefficients of like powers of p, we have

Coefficient of p:

$$y_1' = 1 + y_0^2, \quad y_1(0) = 0$$

Coefficient of p^2:

$$y_2' = 2 y_0 y_1, \quad y_2(0) = 0$$

Coefficient of p^3:

$$y_3' = y_1^2 + 2 y_0 y_2, \quad y_3(0) = 0$$

and so on.

Solving these IVPs, we have

$$y_1(x) = x$$

$$y_2(x) = 0$$

$$y_3(x) = \frac{x^3}{3}$$

and so on.

Hence,

$$y(x) = x + \frac{x^3}{3} + \cdots$$

The exact solution to this problem is $y(x) = \tan(x)$.

Optimal Homotopy Analysis Method

We have seen in HAM that the $\hbar$ curves are used to find the range of the convergence control parameter to identify a convergent series solution to the given problem. But, the major difficulty is that these curves give only a range for the parameter and cannot tell the best value of it that provides the fastest convergent series solution to the problem.

In case of HPM, the convergent control parameter is taken to be –1. It has been observed by several researchers (Liang and Jeffrey, 2009) that giving this value to the parameter did not yield convergent solutions.

Hence, identification of the appropriate convergent control parameter that yields the fastest convergent solution is a problem that needs to be addressed.

In 2007, Yabushita et al. suggested that an optimization method can be used to find out the optimal convergence control parameter. They suggested that by using the least-squares method, which minimizes the sum of the squares of the residuals, the optimum value of the convergence parameter can be evaluated. Marinca and Herisanu (2008) applied this optimization technique and developed the OHAM, which is described in this section.

Optimal Homotopy Asymptotic Method

$$L(u(x)) + N(u(x)) + g(x) = 0, \quad B(u(x)) = 0 \tag{8.19}$$

where L is a linear operator, x is an independent variable, $u(x)$ is an unknown function, $g(x)$ is a known function, N is a nonlinear operator, and $B(u)$ is a boundary operator.

By OHAM, we first construct a family of equations:

$$H(\phi(x), p) = (1 - p)\{L(\phi(x, p)) + g(x)\} - H(p)$$

$$\{L(\phi(x, p)) + g(x) + N(\phi(x, p))\} = 0$$

$$B(\phi(x, p)) = 0 \tag{8.20}$$

where p lies in [0,1] and is an embedding parameter, $H(p)$ is a nonzero auxiliary function for $p \neq 0$ and $H(0) = 0$, and $\phi(x, p)$ is an unknown function.

Obviously, when $p = 0$, $p = 1$, it holds that

$$L(\phi(x, p) + g(x)) = 0$$

whose solution is, say,

$$u_0(x) \tag{8.21}$$

and

$$L(\phi(x, p)) + g(x) + N(\phi(x, p)) = 0$$

which is the given equation and whose solution is the exact solution

$$u(x). \tag{8.22}$$

Thus, as p increases from 0 to 1, the solution $\phi(x, p)$ varies from $u_0(x)$ to the solution $u(x)$.

The auxiliary function $H(p)$ is chosen in the form

$$H(p) = pC_1 + p^2C_2 + \cdots \tag{8.23}$$

where $C_1, C_2 \ldots$ are constants that are to be determined.

Expanding $\phi(x, p)$ in a series with respect to p, we have

$$\phi(x, p, C_i) = u_0(x) + \sum_{k=1}^{\infty} u_k(x, C_i)p^k \quad \text{for} \quad i = 1, 2, \ldots \tag{8.24}$$

Substituting Equation (8.24) in Equation (8.20) and comparing the coefficients of different powers of p, we obtain a set of differential equations with boundary conditions. Solving these differential equations, we obtain $u_0(x), u_1(x, C_1)\ldots$, and the solution of Equation (8.19) can be written as follows:

$$\tilde{u}(x, C_i) = u_0(x) + \sum_{k=1}^{\infty} u_k(x, C_i) \tag{8.25}$$

The residual is calculated as

$$R(x, C_i) = L(\tilde{u}(x, C_i)) + N(\tilde{u}(x, C_i)) + g(x) \tag{8.26}$$

If $R(x, C_i)$ is zero, then $\tilde{u}(x, C_i)$ is the exact solution. Generally, such a case will not arise for a nonlinear problem. Hence, we try to minimize the error occurring in the solution by considering the following functional:

$$J(C_1,C_2,...C_m)=\int_a^b (R(x,C_1,C_2,...C_m))^2\,dx \tag{8.27}$$

where a and b are two values depending on the boundary conditions. The unknown constants $C_i, i=1,2...m$ can be found from the conditions

$$\frac{\partial J}{\partial C_1}=\frac{\partial J}{\partial C_2}=...\frac{\partial J}{\partial C_m}=0 \tag{8.28}$$

Now, with these constants, the approximate solution (of order m) in Equation (8.25) is well determined.

Example 8.6

Solve

$$x^2y''+xy'-4y=0$$

using OHAM.

Solution: Choose

$$L(y)=x^2y''$$

and

$$N(y)=xy'-4y.$$

Now,

$$H(y,p)=(1-p)(x^2y'')+(pC_1+p^2C_2+\cdots)$$
$$(xy'-4)=0$$

Let

$$y(x,p)=y_0+y_1p+y_2p^2+\cdots$$

Substituting y in this homotopy and comparing the coefficients of like powers of p, we have

Coefficient of p:

$$x^2 y_0'' = 0, \quad y_1(1) = 0, \quad y_1'(1) = 2$$

Solving this IVP, we obtain

$$y_0(x) = 2(-1 + x).$$

Substituting $y_0(x) = 2(-1 + x)$ in the constructed homotopy and collecting the

Coefficient of p^2:

$$x^2 y_1' = -(8C_1 - 6C_1 x), \quad y_1(1) = 0, \quad y_1'(1) = 0$$

Again, solving it, we obtain

$$y_1(x) = 2(7C_1 - 7C_1 x + 4C_1 \ln(x) + 3C_1 x \ln(x))$$

Proceeding as previously, we obtain

$$y_2(x) = (2C_1(2 - 7x)) + 80C_1^2(-1 + x) - 7C_2(-1 + x) + ((2C_2(4 + 3x)$$

$$+ C_1(8 + 6x)) + 8C_1^2(11 + 9x))\ln(x) + C_1^2(-16 + 9x)(\ln(x))^2$$

Thus, the solution up to the third approximation is

$$y = y_0 + y_1 + y_2.$$

The residual given by Equation (8.26) is

$$8\text{-}56C_2 + C_1^2(608 - 606x) - 6x + 54C_2 x + 36C_1(-2 + 3x)$$

$$+ 2(-2C_1(16 + 9x) - C_2(16 + 9x) + 2C_1^2(88 + 63x)\ln(x))$$

$$+ C_1^2(64 - 27x)(\ln(x))^2$$

Taking $a = 1$, $b = 1.5$, evaluating the Jacobian as given in Equation (8.27), and using Equation (8.28) for evaluating the constants, we obtain

$$C_1 = 0.001746, \quad C_2 = 0.95266$$

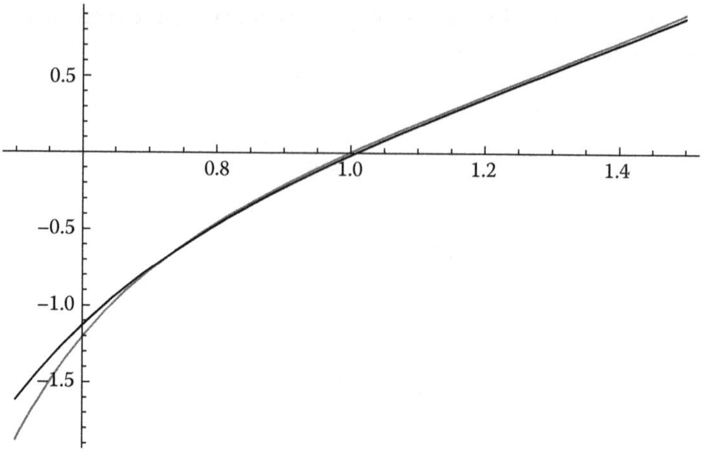

Figure 8.4 Plot of the OHAM (gray) and exact solutions (black).

(Because of the lengthy expressions involved, the final result only is presented.)

Hence, the solution is

$$y = 11.3682 - 11.3856x + (7.64895 + 5.7367x)\ln(x)$$
$$-(0.0000487942 + 0.0000274467x)(\ln(x))^2$$

The exact solution to the problem is (Figure 8.4)

$$y = \frac{1}{2}(x^2 - x^{-2})$$

Example 8.7

Solve

$$\frac{d^2v}{dr^2}\left(1 - \frac{dv}{dr}\right) + \frac{1}{r}\frac{dv}{dr}\left(1 - \frac{1}{2}\frac{dv}{dr}\right) + 1 = 0$$

with

$$v(0.2) = 0, \quad v(1) = 0$$

using OHAM.

Solution: Choose

$$L(v)=\frac{d^2v}{dr^2}+1$$

and

$$N(v)=\frac{d^2v}{dr^2}\left(-\frac{dv}{dr}\right)+\frac{1}{r}\frac{dv}{dr}\left(1-\frac{1}{2}\frac{dv}{dr}\right).$$

Now,

$$H(v,p)=(1-p)\left(\frac{d^2v}{dr^2}+1\right)+(pC_1+p^2C_2+\cdots)$$

$$\left(\frac{d^2v}{dr^2}\left(-\frac{dv}{dr}\right)+\frac{1}{r}\frac{dv}{dr}\left(1-\frac{1}{2}\frac{dv}{dr}\right)\right)=0$$

Let

$$v(x,p)=v_0+v_1p+v_2p^2+\cdots$$

Substituting y in this homotopy and comparing the coefficients of like powers of p, we have

Coefficient of p:

$$v_0''=-1,\quad v_0(0.2)=0,\quad v_0(1)=0$$

Solving this IVP, we obtain

$$v_0(x)=\frac{(-0.2+1.2x-x^2)}{2}.$$

Substituting $v_0(x)$ in the constructed homotopy and collecting the

Coefficient of p^2:

$$v_1''=0.2C_1+\frac{0.42c_1}{x}-1.5C_1x,\quad v_1(0.2)=0,\quad v_1(1)=0$$

Again, solving it, we obtain

$$v_1(x)=C_1(0.1289+0.021x+0.1x^2-0.25x^3+0.42x\log(x))$$

Proceeding as previously, we can find an expression for $v_2(x)$ that will be in terms of C_1 and C_2.

Thus, the solution up to the third approximation is

$$v = v_0 + v_1 + v_2.$$

Finding the residual and evaluating the Jacobian as given in Equation (8.27), we obtain

$$C_1 = -0.9795, \quad C_2 = 0.9937$$

(Because of the lengthy expressions involved, the final result only is presented. You may check using Mathematica® if you are unconvinced.)

Hence, the solution is

$$v(x) = -0.3515 + 0.7537x - 0.5955x^2 + 0.4332x^3 - 0.2398x^4$$
$$+ (-0.6391 + 0.4029x)x\log(x) + 0.0805x(\log(x))^2$$

Exercise Problems

1. Solve the boundary value problem

$$u'' + \varepsilon(u - u^3) = 0, \quad u(0) = u(\pi) = 0$$

Hint: Take

$$u_0(x) = \sin(x); \ L \equiv u'' + u; \ H(x) = 1$$

2. Solve

$$u'' + u^2 = 1, \quad u(0) = 0$$

Hint: See Liao's work (2003), in which this problem has been solved by different methods illustrated in this text.

Applications

S. Abbasbandy. The application of homotopy analysis method to nonlinear equations arising in heat transfer. *Physics Letters A*, Vol. 360, pp. 109–113, 2006.

M. Ayub, A. Rasheed, and T. Hayat. Exact flow of a third grade fluid past a porous plate using homotopy analysis method. *International Journal of Engineering Science*, Vol. 41, pp. 2091–2103, 2003.

M. Babaelahi, G. Domairry, and A.A. Joneidi. Viscoelastic and MHD flow boundary layer over a stretching surface with viscous and ohmic dissipations. *Meccanica*, Vol. 45, pp. 817–827, 2010.

M.S.H. Chowdhury, I. Hashim, and O. Abdulaziz. Comparison of HAM and HPM for purely nonlinear fin-type problems. *Communications in Nonlinear Science and Numerical Simulation*, Vol. 14, pp. 371–378, 2009.

E. Cuce and P.M. Cuce. Homotopy perturbation method for temperature distribution, fin efficiency and fin effectiveness of convective straight fins with temperature-dependent thermal conductivity. *Journal of Mechanical Engineering Science*, Vol. 228, No. 8, 2012.

D.D. Ganji. The application of He's homotopy perturbation method to nonlinear equations arising in heat transfer. *Physics Letters A*, Vol. 355, pp. 337–341, 2006.

D.D. Ganji, H. Babazadeh, F. Noori, M.M. Pirouz, and M. Janipou. An application of homotopy perturbation method for non-linear blasius equation to boundary layer flow over a flat plate. *International Journal of Nonlinear Science*, Vol. 7, pp. 399–404, 2009.

D.D. Ganji and N. Jamshidi. Application of He's homotopy-perturbation method to nonlinear coupled systems of reaction-diffusion equations. *International Journal of Nonlinear Science and Numerical Simulation*, Vol. 7, pp. 413–420, 2006.

D.D. Ganji and A. Sadighi. Application of homotopy-perturbation and variational iteration methods to nonlinear heat transfer and porous media equations. *Journal of Computational and Applied Mathematics*, Vol. 207, pp. 24–34, 2007.

T. Hayat, M. Khan, and S. Asghar. Homotopy analysis of MHD flows of an Oldroyd 8-constant fluid. *Acta Mechanica*, Vol. 168, pp. 213–232, 2004.

J.H. He. Homotopy perturbation method for bifurcation of nonlinear problems. *International Journal of Nonlinear Sciences and Numerical Simulation*, Vol. 6, pp. 201–206, 2005.

J.H. He. A review on some new recently developed non linear analytical techniques. *International Journal of Nonlinear Sciiences and Numerical Simulation*, Vol. 1, pp. 51–70, 2000.

E.O. Ifidon. An application of homotopy analysis to the viscous flow past a circular cylinder. *Journal of Applied Mathematics*, Vol. 2009, 52304, 2009.

S. Iqbal, A.R. Ansari, A.M. Siddiqui, and A. Javed. Use of optimal homotopy asymptotic method and Galerkin's finite element formulation in the study of heat transfer flow of a third grade fluid between parallel plates. *ASME, Journal of Heat Transfer*, Vol. 133, 091702, 2011.

J.I. Hossein, M. Zabihi, and M. Saidy. Application of homotopy-perturbation method for solving gas dynamics equation. *Applied Mathematical Sciences*, Vol. 48, pp. 2393–2396, 2008.

M. Khaki and D.D. Ganji. Analytical solutions of nano boundary layer flows by using He's perturbation method. *Mathematical and Computational Applications*, Vol. 15, pp. 962–966, 2010.

S.J. Liao. The Proposed Homotopy Analysis Technique for the Solution of Nonlinear Problems. PhD thesis, Shanghai Jiao Tong University, 1992.

V. Marica and N. Hersanu. Application of optimal homotopy asymptotic method for solving nonlinear equations arising in heat transfer. *International Journal of Heat and Mass Transfer*, Vol. 13, pp. 710–715, 2008.

R. Nawaz, M.N. Khalid, S. Islam, and S. Yasin. Solution of tenth order boundary value problems using optimal homotopy asymptotic method (OHAM). *Canadian Journal on Computing in Mathematics, Natural Sciences, Engineering & Medicine,* 1, pp. 37–54, 2010.

T.S.L. Radhika and A.V. Singh. Application of homotopy asymptotic method for non Newtonian fluid flow through a vertical annulus. *Proceedings International Multiconference of Engineers and Computer Scientists*, Vol. 2, pp. 1384–1389, 2012.

R.A. Shah, S. Islam, and A.M. Siddiqui. Couette and Poiseuille flows for fourth order fluids using OHAM. *Applied Sciences Journal,* Vol. 9, pp. 1228–1236, 2010.

J. Saberi-Nadjafi and A. Ghorbani. He's homotopy perturbation method: An effective tool for solving nonlinear integral and integro-differential equations. *Computers and Mathematics with Applications*, Vol. 58, pp. 2379–2390, 2009.

H. Saberinik, M.S. Zahedi, R. Buzhabadi, and S. Effati. Homotopy perturbation method and He's polynomials for solving the porous media equation. *Computational Mathematics and Modeling*, Vol. 24, pp. 279–292, 2013.

I. Shafieenejad, N. Moallemi, H.H. Afshari, and A.B. Novinzadeh. Application of He's homotopy perturbation method for pipe flow of non-Newtonian fluid. *Advanced Studies in Theoretical Physics*, Vol. 3, pp. 199–211, 2009.

Bibliography

J.-H. He. An approximate solution technique depending upon an artificial parameter. *Communications in Nonlinear Sciences and Numerical Simulation*, Vol. 3, No. 2, 92–97, 1998.

J.H. He. Homotopy perturbation technique. *Computer Methods in Applied Mechanics and Engineering*, Vol. 178, pp. 257–262, 1999.

J.H. He. A coupling method of a homotopy technique and a perturbation technique for non-linear problems. *International Journal of Nonlinear Mechanics*, Vol. 35, pp. 37–43, 2000a.

J.H. He. Homotopy perturbation method: A new nonlinear analytical technique. *Journal of Applied Mathematics and Computing*, Vol. 135, pp. 73–79, 2000b.

J.H. He. Homotopy perturbation method: A new non-linear analytical technique. *Applied Mathematics and Computation*, Vol. 135, pp. 73–79, 2003.

J.H. He. Comparison of homotopy perturbation method and homotopy analysis method. *Applied Mathematics and Computation*, Vol. 156, pp. 527–539, 2004.

S.X. Liang and D.J. Jeffrey. Comparison of homotopy analysis method and homotopy perturbation method through an evolution equations. *Communications in Nonlinear Science and Numerical Simulation*, Vol. 14, pp. 4057–4064, 2009.

S.J. Liao. *Beyond Perturbation: Introduction to the Homotopy Analysis Method.* Boca Raton, FL: Chapman & Hall/CRC Press, 2003.

S. Liao. Comparison of homotopy analysis method and homotopy perturbation method. *Applied Mathematics and Computation*, Vol. 169, pp. 1186–1194, 2005.

S.J. Liao. Notes on the homotopy analysis method: Some definitions and theorems. *Communications in Nonlinear Science and Numerical Simulation,* Vol. 14, pp. 983–997, 2009.

V. Marinca and N. Herisanu. Application of optimal asymptotic method for solving nonlinear equations arising in heat transfer. *International Communications in Heat and Mass Transfer*, Vol. 35, pp. 710–715, 2008.

K. Yabushita, M. Yamashita, and K. Tsuboi. An analytical solution of projectile motion with the quadratic resistance law using the homotopy analysis method. *Journal of Physics A, Mathematical and Theoretical*, Vol. 40, pp. 8403–8416, 2007.

Index

E

F

G

H

R

Milton Keynes UK
Ingram Content Group UK Ltd.
UKHW040056071024
449327UK00019B/607

9 780367 378127